SpringerBriefs in Molecular Science

SpringerBriefs in Molecular Science present concise summaries of cutting-edge research and practical applications across a wide spectrum of fields centered around chemistry. Featuring compact volumes of 50 to 125 pages, the series covers a range of content from professional to academic. Typical topics might include:

- A timely report of state-of-the-art analytical techniques
- A bridge between new research results, as published in journal articles, and a contextual literature review
- A snapshot of a hot or emerging topic
- An in-depth case study
- A presentation of core concepts that students must understand in order to make independent contributions

Briefs allow authors to present their ideas and readers to absorb them with minimal time investment. Briefs will be published as part of Springer's eBook collection, with millions of users worldwide. In addition, Briefs will be available for individual print and electronic purchase. Briefs are characterized by fast, global electronic dissemination, standard publishing contracts, easy-to-use manuscript preparation and formatting guidelines, and expedited production schedules. Both solicited and unsolicited manuscripts are considered for publication in this series.

Pratima Bajpai

Developments in Microbial Bioremediation

A Guide for Sustainable Waste Treatment

 Springer

Pratima Bajpai
Consultant (Pulp and Paper)
Kanpur, Uttar Pradesh, India

ISSN 2191-5407   ISSN 2191-5415 (electronic)
SpringerBriefs in Molecular Science
ISBN 978-3-031-78318-0   ISBN 978-3-031-78319-7 (eBook)
https://doi.org/10.1007/978-3-031-78319-7

This Springer imprint is published by the registered company Springer Nature Switzerland AG
The registered company address is: Gewerbestrasse 11, 6330 Cham, Switzerland

If disposing of this product, please recycle the paper.

Preface

The problem of various types of environmental degradation is currently being faced by the entire world. Microorganisms are required for an essential contingency strategy to overcome barriers. Because of their extraordinarily high metabolic activity, microorganisms may thrive in a variety of environments and habitats throughout the biosphere. Their extensive range of nutritional capabilities allows for their application in the bioremediation of environmental pollution. Bioremediation relies on the ubiquitous activity of microorganisms to facilitate the degradation, removal, immobilization, or detoxification of numerous chemical wastes and hazardous physical substances from the environment. The core principle involves the breakdown and transformation of pollutants such as oil, heavy metals, pesticides, dyes, and hydrocarbons. This process plays a crucial role in addressing numerous environmental challenges, as it occurs enzymatically through metabolic processes. The rate of degradation is influenced by two primary categories of factors: biotic and abiotic conditions. Various methodologies and strategies are currently employed across different regions globally. Each bioremediation technology presents distinct advantages and disadvantages depending on its specific application. This book offers advanced technologies, current insights, and future perspectives for scientists and researchers engaged in wastewater treatment, effluent treatment facilities, and the biodegradation of environmental pollutants, all aimed at promoting environmental safety and sustainable development.

Kanpur, India   Pratima Bajpai

Acknowledgments I express my sincere gratitude to the numerous individuals and organizations that have contributed valuable information. I also wish to extend my thanks to the various publishers who granted me permission to utilize their content. My deepest appreciation goes to Elsevier, Springer, Hindawi, MDPI, IntechOpen, Frontiers, SpringerOpen, and other open-access journals and publications.

Contents

List of Figures

List of Tables

Chapter 1
Introduction and General Background of Microorganisms in Waste Management

Abstract Population growth, industrialization, and urbanization are causing environmental issues like waste generation, pollution, and chemical contaminants. Waste management strategies involve sorting, collecting, transporting, recycling, and disposing of waste. Microbial technologies are increasingly used in waste management and bioremediation to maintain environmental balance and neutralize pollutants. An introduction and a comprehensive overview of the role of microorganisms in waste management are given in this chapter.

Keywords Microorganisms · Waste management · Bioremediation · Biodegradation · Microbial detoxification · Phytoremediation · Biotransformation · Environmental sustainability · Landfills · Incineration · Composting · Gasification

1.1 General Background

Due to the tremendous strain that an expanding population, industrialization, and fast urbanization are placing on natural resources, the current status of our global ecosystem is beset with serious problems. Pollution is caused by the enormous volumes of garbage produced by non-scientific methods in modern society (Raj et al. 2018a, b). Moreover, big-scale businesses' extensive consumption of raw materials results in the irreversible release of vast amounts of chemical pollutants and radioactive waste into the environment, endangering the biosphere. Trash generation is the primary source of material and energy loss. It also has detrimental impacts on the environment and drives up the expenses associated with waste management, which includes collection, treatment, disposal, and general management (Jhariya et al. 2018; Mondal and Palit 2019).

In order to avoid a predicted 70% rise in global waste by 2050, immediate action is required, according to the recent report by World Bank—What a garbage 2.0: A Global Snapshot of Solid Waste Management to 2050. The estimate states that growing urbanization and population growth will cause the quantity of waste produced globally to increase to 3.4 billion tons over the next 30 years from 2.01 billion tons in 2016.

About 34% of the world's waste is produced in higher income countries, but making up only 16% of the worldwide population.

Approximately 25% of the world's waste is produced in the Pacific and East Asian regions. By the year 2050, the amount of waste produced in Sub-Saharan Africa is expected to increase by over three times, whereas waste in South Asia is anticipated to rise by over four times.

The production of toxic waste varies widely among cities and is closely linked to urban expansion. Without the implementation of scientific waste management practices, it is anticipated that waste generation will rise in parallel with industrialization rates (Sharma and Shah 2005).

Waste materials or substances can be categorized as solid, liquid, or gaseous. They are commonly emitted as industrial wastewaters, sewage sludge, electronic scraps, nuclear waste, and household garbage (Hassan et al. 2003; Angulo et al. 2010). Once disposed of, these pollutants often undergo transformation into new poisonous or hazardous forms over time.

Solid waste is a broad category that includes many different waste kinds, including regular home garbage, sewage sludge, industrial waste, waste from construction and demolition, agricultural waste, mining waste, waste from food processing, and waste from petroleum extraction (Demirbas 2011). Liquid waste comprises spent oils, stormwater, residential wastewater (from kitchens, baths, and laundry rooms), and wastewater from industrial processes. Tiny particles and fumes from automobiles, open fires, incinerators, and industrial and agricultural activities make up gaseous waste. Once released into the environment, it becomes impossible to regulate and eliminate these gases and particles (Hassan et al. 2003; Chowdhary et al. 2018).

Due to the wide-ranging sources stemming from our daily activities, there are numerous types of waste. The composition of these diverse wastes, resulting from emerging sectors and ongoing technological advancements, varies over time and space, making them hazardous waste items. The strain that these industrial wastes impose on marine, terrestrial, and atmospheric ecosystems has led to their categorization as hazardous materials.

The nation's wealth and overall progress, along with the responsible use and accumulation of various resources, are directly connected to the growth of the human population (Weltens et al. 2012). It is absolutely essential to eliminate any harmful and toxic compounds from sludge or wastewater released into the environment for the sake of protecting the ecosystem and public health.

Waste is classified based on its source, including households, businesses, hospitals, farms, and other establishments, as well as its characteristics such as solid, liquid, gas, or waste heat (Fig. 1.1).

Waste management scenarios differ significantly between developed and developing countries; underdeveloped countries often lack proper garbage collection and disposal infrastructure. Given the increasing concerns about environmental degradation and sustainability, waste management has grown to be a very significant issue. On the outskirts of towns and villages, careless, unmanaged, and random trash

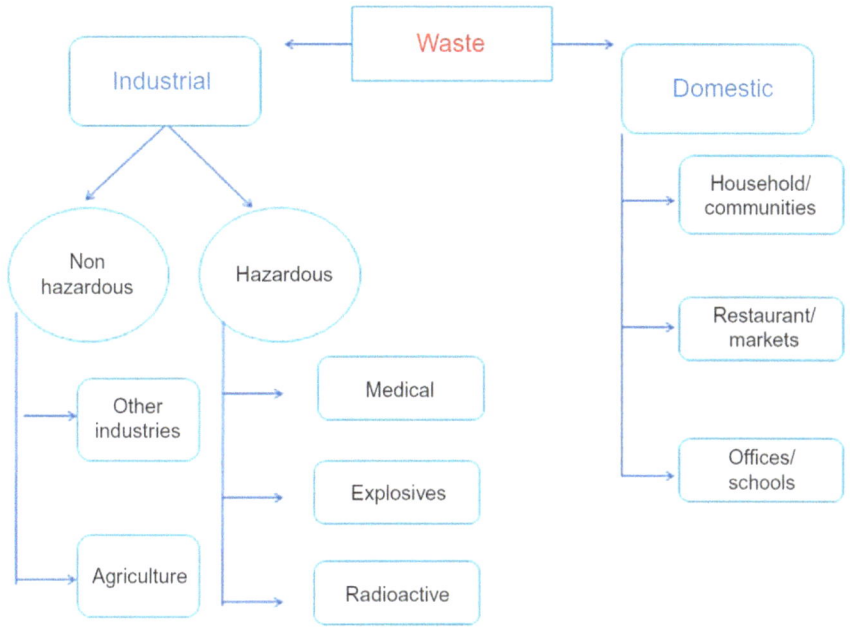

Fig. 1.1 Classification of waste. Reproduced with permission from Mani et al. (2020)

dumping results in overflowing landfills, which pose major environmental risks by contaminating soil and groundwater and making it impossible to return the landfills to a suitable state. An operable waste management system that is both economically and environmentally sustainable is necessary given the current circumstances.

Numerous commonly employed methods exist for the management and treatment of trash, including:

1. Incineration method: involves high-temperature treatment of wastes and rubbish items.
2. Sanitary landfills a more beneficial and widely used approach where refuse is stored away from natural surroundings.
3. Recycling: the process of changing resources for new uses.
4. Avoidance and minimization: avoiding the purchase of new items and instead fixing broken ones, using reusable products, and reusing second-hand goods.

Therefore, sustainable cleanup of contaminated environments is crucial. In the recent years, there has been an increase in the importance of microorganisms in the biodegradation of pollutants and waste management (Banerjee et al. 2018). Many microorganism-based biotechnological methods, including composting, biotransformation, biodegradation, and bioremediation, have been effectively used to accumulate and break down contaminants (Mondal and Palit 2019; Hussain and Dhanker 2021; Husain et al. 2022; Parmar et al. 2021; Sharma et al. 2023; Mani et al. 2020). According to Maghraby and Hassan (2018), green algae, or *Cladophora* sp., can serve as a powerful and effective alternative for waste management due to its high bioaccumulation propensity for hazardous metals. Furthermore, the effective operation of biological processes in wastewater treatment systems is greatly dependent on microbial ecology.

Martinez et al. (2018), demonstrated that bacteria and archaea are the primary inhabitants of the bioreactors in the wastewater treatment system located in the polar arctic region of Finland. A process referred to as "nanobioremediation" has been employed to effectively enhance microbial activity through the use of nanoparticles. Additionally, the extremophilic bacterium *Deinococcus radiodurans*, known for its natural resistance to radiation, is utilized for the disposal of radioactive waste in the USA (Varma et al. 2017; Brim et al. 2000). Consequently, the most efficient method for treating various types of waste involves the utilization of microorganisms in conjunction with various biotechnologies, which not only proves to be economically viable but also environmentally sustainable and eco-friendly.

The careless dumping of rubbish adds to the pollution issue and demonstrates the effects of human activity on the environment (Table 1.1). This leads to the release of contaminants into the ecosystem, causing significant depletion. The expansion of industrialization, healthcare facilities, and agricultural activities results in the generation of substantial amounts of agricultural, industrial waste, and biomedical waste, all of which are harmful to the health of the people and the environment.

Waste can be broadly categorized into gaseous, liquid, and solid types (Hussain and Dhanker 2021) (Table 1.2), emanating from sources such as municipal, industrial, biomedical, and electronic.

Following a methodical approach that includes the appropriate collection, storage, and scientific disposal of waste materials is essential for addressing waste management.

Reducing the negative effects of waste products is the main goal of waste management. The four R's—refuse, reduce, reuse, and recycle—capsulate this. A wide range of tasks are included in waste management, such as waste creation, handling, collection, and transportation as well as recycling, treatment, and disposal.

Bioremediation serves as a valuable process to either completely eliminate contaminants or significantly reduce their concentration (Alexander 1994; Gu et al. 2000) Fungi, bacteria and plants play an important role in this process falling under the categories of myco-remediation, remediation, and phytoremediation. These

Table 1.1 Process of waste generation

Human activity (industrialization, agriculture processing, medical processing, urbanization)
Waste generation solid liquid and gas
Storage of waste
Collection of waste
Disposal of waste

Based on Husain et al. (2022)

Table 1.2 Waste categories

Matter based
Solid
Liquid
Gas
Source based
Household
Industrial
Radioactive
Electronic
Medical
Nuclear
Degradation feature based
Biodegradable
Non-biodegradable
Environmental impact based
Hazardous
Nonhazardous

Based Husain et al. (2022)

processes incorporate a range of sub-processes such as biological stimulation, biological augmentation, and biological sparging to further enhance on-site toxin removal. While biostimulation rejuvenates native microorganisms by providing them with additional nourishment and growth ingredients, bioaugmentation involves introducing foreign bacteria to an ecosystem to degrade contaminants. Additionally, biosparging employs compressed air to supply nutrients and oxygen to specific areas, thereby stimulating microbial activity.

The increasing challenges posed by environmental issues are a growing concern for humanity, particularly due to the expanding industrialization and globalization. Industrial wastewater presents several key features such as high levels of total dissolved solids, excessive alkalinity, and significant biological and chemical oxygen demand. Additionally, it contains heavy metals and effluents from industries like textiles, which pose a considerable environmental risk due to the use of a variety of colors and auxiliary chemicals in the production of high-quality products. Wastewater treatment, particularly through biotechnological processes, is of paramount importance globally due to the exceptional abilities demonstrated by microbial communities in degrading a wide spectrum of pollutants from both manmade and natural systems.

Microbes play a crucial role in eliminating diverse pollutants from various biological wastewater treatment systems. The utilization of fungi in wastewater cleanup offers several benefits, such as the production of fungal biomass for human and animal consumption, and the transformation of organic waste materials into valuable fungal protein and biochemicals. Microbial activity has also been observed to alter the oxidation state of heavy metals. Various effective bioremediation techniques have been developed for cleaning up contaminated areas, with the potential for enhanced bioremediation through the strategic use of genetically modified microorganisms

(GEM). This approach allows for the creation of designer biocatalysts capable of effectively degrading a variety of pollutants, including resistant substances.

1.2 Types and Origins of Waste

The term "wastes" refers to materials that are thrown away after their intended useful life. Waste management involves various activities such as sorting, collecting, transporting, recycling, and disposing of different types of wastes.

The rapid industrialization driven by efforts to boost economic development and improve living standards has caused a notable rise in the quantity of waste produced every day (Parmar et al. 2021; Nandan et al. 2017; Giusti 2009; Husain et al. 2022; Mani et al. 2020; Sharma et al. 2023; Ochedi et al. 2020; Mondal and Palit 2019; Parmar et al. 2021).

1.2.1 Waste Classifications and Their Sources

A major contributor to environmental pollution is the waste produced by various manufacturing processes, industries, and municipal sources, with industries being the primary culprit. Solid waste is a general term for the waste produced by these industries. This contributes to soil pollution when added to landfills, and liquid waste, which poses a significant threat to the environment, particularly aquatic ecosystems (Holt 2001). Groundwater contamination is often caused by waste disposal from municipal, industrial, and military sources, as well as activities such as mining, chemical spills, and pesticide use.

Certain industries contribute specific types of waste that present unique challenges for treatment and disposal. For example, textile industries produce effluents containing complex organic compounds, pharmaceutical industries generate acidic effluents with high chemical oxygen demand and organic solvents, and wine industries produce acidic effluents rich in organic fractions such as sugar, alcohol, and polyphenols. Agricultural activities also generate waste, including crop residue, straws, husks, and various byproducts. Refer to Table 1.3 for a comprehensive summary of various waste categories. Figure 1.2 illustrates the origins of waste generation (Mondal and Palit 2019). For a broad classification of prominent waste sources, consult Fig. 1.3 (Parmar et al. 2021).

Overall, the increasing volume and diverse nature of waste generated by different sectors pose significant challenges to effective waste management and environmental protection. Various industries produce a wide range of pollutants during the production of goods. These pollutants can be categorized as organic or inorganic. Organic pollutants include substances like phenols, esters, azo dyes, hydrocarbons, petroleum, pesticides, and persistent organic pollutants. On the other hand, inorganic pollutants,

Table 1.3 Waste classification

Solid Waste Food remnants, rotting produce, agricultural waste, cans, bottles, metal fragments, plastics, wrappers, ashes, and dead animal parts are all included in this category
Liquid Waste This covers oil spills as well as sewage from restrooms in homes, workplaces, hospitals, restaurants, and offices
Gaseous Waste This group includes fuel exhausts that include smog, carbon dioxide, carbon monoxide, nitrogen oxides, and sulfur dioxide
Biodegradable Waste Simple Biodegradable Waste: Over time, these waste materials easily break down due to natural decomposition processes. This group includes leaves, vegetable peels, plant remains, fecal matter, wastewater, dead plants, and animals Complex Biodegradable Waste: This type of garbage resists natural decomposition processes and does not break down easily. That being said, it can decompose over time
Municipal Waste: Waste generated inside a municipality or local government unit is referred to as municipal waste. This includes garbage that is collected from public trash cans and comes from a variety of sources, including stores, restaurants, banks, courts, schools, libraries, hospitals, and parks
Biomedical Waste Biomedical wastes are produced as a result of diagnosing, treating, and immunizing human beings or animals. This group also includes trash from animal research and experiments, as well as microbiological waste from human and animal cell cultures, microorganisms, and laboratory cultures
Commercial Waste/Business Waste Waste from diverse small and large-scale industries is termed as such. Industrial waste involves any material rendered useless during production. Chemicals (arsenic, mercury, lead, etc.), paints, paper products, sandpaper, metals, and industrial byproducts are a few examples. Industries generate waste gases and other materials in addition to using a lot of fuel for energy. The emissions from power stations, specifically sulfur dioxide and nitrogen oxides, lead to the development of acid rain and pose significant health risks.
Nuclear Waste Radioactive waste produced from the nuclear energy industry, including substances used in cooling and storing nuclear fuel, nuclear fuel from reactors in power stations and submarines, and X-ray machines located in hospitals and airports. Radioactive elements like uranium and radium have very unstable atomic nuclei, emitting radiation that can be extremely harmful
Electronic Waste This term encompasses outdated or no longer useful electronic objects such as computer monitors, TVs, VCRs, stereos, mobile phones, and other functional or damaged gadgets

such as nickel, lead, chromium, and mercury are more harmful and resistant to degradation. Industries are the main cause of environmental pollution, although homes, farms, livestock, and landfills also have a big impact (Saxena and Bhargava 2017). For example, household waste comprises of organic waste like vegetable and fruit

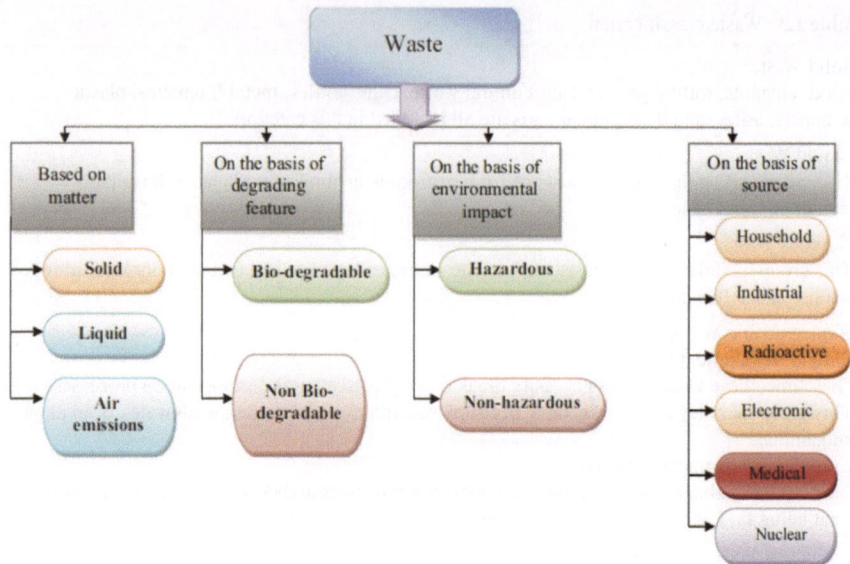

Fig. 1.2 Different types of waste. Reproduced with permission from Mondal and Palit (2019)

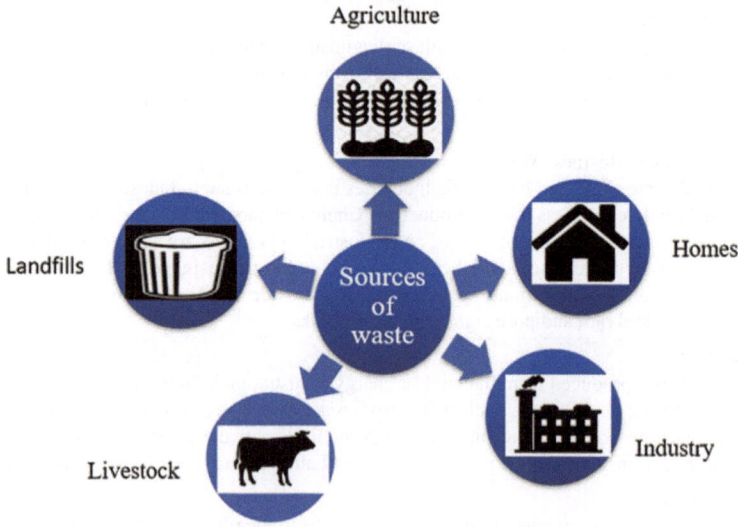

Fig. 1.3 Sources of waste. Reproduced with permission from Parmar et al. (2021)

leftovers, dust, detergent drains, and sewage. Livestock waste mainly includes animal waste, bedding material, contaminated soils, hair, feathers, and other debris.

The goal of waste management is to reduce the negative impact that waste materials have on the environment and human health by organizing, collecting, disposing of, and storing waste materials. Waste management is becoming more and more difficult as a result of the industrial revolution, strain from the population on natural resources and increased urbanization. It comprises four main categories of waste—industrial, electronic, municipal, and biomedical—each governed by specific policies. The fundamental principles of waste management are encapsulated in the 4R theory: refuse, reduce, reuse, and recycle. The main focuses of India's waste management plan include trash generation, disposal, recycling, treatment, and storage as well as collection and transportation. Common waste management methods include landfills, incineration, composting, and gasification (Yadav et al. 2023). For a visual representation of the waste management system, refer to Fig. 1.4 (Mondal and Palit 2019).

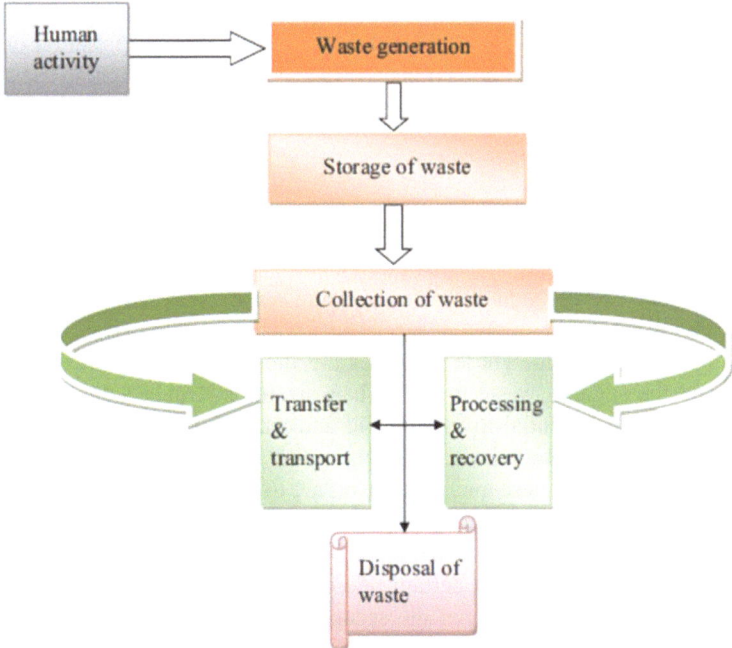

Fig. 1.4 Schematic view of waste management system. Reproduced with permission from Mondal and Palit (2019)

1.3   The Importance of Microorganisms in Waste Management

Microorganisms play a crucial role in maintaining environmental balance by being associated with various natural and human-made phenomena (Wu et al. 2020). They are able to thrive in diverse environmental conditions, including harsh ones and are responsible for recycling natural materials. Microbes have been shown to have benefits in genetic engineering, industrial wastewater treatment, environmental protection, and human health (Parmar et al. 2021; Abatenh et al. 2017; Daniel et al. 2022; Satyanarayana et al. 2012; Barea et al. 2005). There is a growing need for microbial solutions to tackle environmental issues like waste management and bioremediation. Commonly employed for waste management, algae, fungi, bacteria, and yeasts, contribute to the natural cycles of carbon and nitrogen. Microbes participate in the bioremediation process by utilizing pollutants or waste as a carbon source for producing energy and byproducts like carbon dioxide and water. Additionally, the use of microorganisms in important environmental domains like sewage treatment is crucial. Sewage, a complex mixture of domestic and industrial wastes, requires accurate characterization for effective treatment and to prevent pollution of water bodies. Understanding the composition of sewage is crucial for designing appropriate processing plants and establishing cost-effective waste management systems.

1.3.1   Sewage Treatment

The presence of microorganisms and parasites in sewage water is a common occurrence. Upon reaching the treatment plant, the sewage typically carries between 10^5 and 10^6 microbes per ml. These microbes are mostly derived from two sources: sewage and the sanitary waste in the soil. A sizable microbial community found in these sources is essential to the aerobic and anaerobic degradation of organic compounds. The most common microorganisms in sewage water are viruses, bacteria, and protozoa. These microorganisms are transported with the sewage water when it is treated. However, depending on where the sewage comes from, these microbes' type and amount can change.

It is interesting to learn that certain microbes help remove certain toxins from sewage, which is a critical part of the secondary treatment process. Furthermore, certain other microorganisms contribute to the tertiary treatment, aiding in the elimination of pollutants like phosphorus and nitrogen. Most microorganisms are removed from water during the disinfection process by ozonation and chlorination.

1.3.1.1   Bacteria

When helpful and hazardous bacteria come into touch with sewage water, they combine to form a mixture in the sewage treatment vats.

- Beneficial Bacteria:

  During the trickling filter phase of secondary treatment, certain biofilm-forming bacteria, such as *Flavobacterium, Chromobacter, Pseudomonas,* and *Zooglea*, are present and aid in the removal of organic material from sewage. Sewage also contains bacteria that remove nitrogen, such as *Thiobacillus* (denitrifier), *Desulfotomaculum* (sulfate reducer), *Nitrosomonas* and *Nitrobacter* (nitrifiers), and *Desulfovibrio* (sulfate reducer).
- Harmful Bacteria:

  Certain bacteria found in sewage may lead to a variety of disorders in humans, pets, and plants when they are consumed through contaminated food or water, through the fecal–oral route, etc. Waterborne illnesses are caused by human pathogens, including *Salmonella typhi* (which causes typhoid), *Vibrio cholera* (which causes cholera), and *Escherichia coli* serotype O157:H7 (which causes food poisoning and diarrhea). Fecal indicators are made with *E. coli.* Most of these microbes are successfully removed and flushed out of the sewage by the disinfection process.

1.3.1.2   Fungi

In fixed film techniques, fungi—both single- and multicellular eukaryotic organisms—help remove carbonaceous elements from sewage.

1.3.1.3   Protozoa

Throughout the sewage treatment process, various types of unicellular eukaryotic organisms like amoebae, flagellates, and ciliates are found. Their primary role involves feeding on organic particulate matter and helping to maintain a slime layer in trickling filter systems. Additionally, these protozoa also assist in controlling bacterial population density through their predatory activities. Protozoa are helpful in treating sewage, but if consumed by humans, they can be dangerous. For instance, the waterborne pathogen *Giardia lamblia*, a protozoan parasite prevalent in sewage, causes giardiasis when its cysts contaminate the water supply with feces.

1.3.1.4   Viruses

More than 140 different species of enteric viruses can pollute sewage. These viruses can enter the human body through the fecal–oral route. Once within the body, the viruses begin to grow in the gastrointestinal system before being expelled in enormous amounts from the infected person's feces. Human enteric viruses commonly found in sewage include enteroviruses like serotype 3 of the poliovirus, which causes paralysis and aseptic meningitis; serotype 23 of the coxsackievirus, which causes aseptic meningitis, paralysis, and Herpangia; serotype 6 of the coxsackievirus, which causes aseptic meningitis, Pleurodynia, and myocarditis; and serotype 34 of the echovirus, which causes respiratory infections, aseptic meningitis, and diarrhea.

Furthermore, rotaviruses (serotype 4) cause gastroenteritis, adenoviruses (serotype 41) cause respiratory infections, gastroenteritis, and acute conjunctivitis, while norwalk agent (serotype 1) and asterviruses (serotype 5) cause gastroenteritis. Hepatitis A virus also causes infectious hepatitis. Thankfully, disinfection is a method used in sewage treatment that eliminates these viral viruses.

1.3.1.5   Algae and Cyanobacteria

Sewage doesn't just contain viral pathogens, but also algae and cyanobacteria like *Euglena, Chlorella, Anacystis, Oscillatoria,* and *Stigeoclonium.* These microorganisms form a biofilm on exposed surfaces and can hamper the trickling process by blocking filters, even though they play a minimal role in fixed film processes.

1.3.1.6   Metazoa

Metazoa—multicellular eukaryotic organisms—such as worms (nematodes, rotifers, oligochaetes, and gastrotricha) and arthropods (crustaceans, tardigrades, and insects) are commonly found in trickling filter systems where they devour the biofilm. These predatory and detrivorous organisms are typically found in both aerobic and microaerophilic conditions in old biomass.

1.3.2   Energy Production

The use of microorganisms for production of energy has been the subject of much research recently in a number of areas, incorporating biochemical processes, such as microbial fuel cells, and environmental nanotechnology. Microbes are used in biogas reactors to produce methane and in fermentation to produce ethanol. Additionally, scientists are investigating different bacteria and algae to produce biofuel from agricultural waste. The potential uses of microbes extend to sewage treatment plants, where they could be employed for simultaneous waste digestion and energy generation to power different facilities (Inslee and Hendricks, 2009; Moradian et al. 2021).

Microbes are a direct source of bioenergy as well as a help in converting biomass into biofuels (Venkatesh and Krishnaveni 2020; Narwal et al. 2020; Dunlop 2011; Singh Pandya et al. 2023). Microalgae and oleaginous bacteria, for instance, store intracellular oil that can be utilized to produce biofuel. Furthermore, several microorganisms emit useful substances that may function as indirect sources of biofuels. Microbial fuel cells use microorganisms to produce bioelectricity and biohydrogen from chemical molecules. Oligomeric microbes are microorganisms that have the ability to retain oil within their cells, typically making up 20% of their total biomass (Meng et al. 2009; Matuszewska 2016).

Microbial oil is also produced by particular species of fungi, yeast, bacteria, and microalgae. Prokaryotic bacteria make specific lipids, whereas eukaryotic fungus, yeast, and microalgae produce triacylglycerols, which can be transesterified into biodiesel.

Microbes use their metabolism to produce useful products for the production of biofuels, converting the chemical energy of biomass into fuels with energy. The successful synthesis of biofuels is thought to depend on the choice of microorganisms, substrate, and manufacturing method. It is beneficial to produce biofuels with a better positive net energy balance in order to reach commercial viability.

The use of metabolic engineering has proven to be a game-changer in the generation of biofuels. In order to produce biofuel, microbes have unique metabolic pathways and enzymes. By manipulating these processes, the yield of microbial biofuel production can be greatly increased. Furthermore, one crucial tactic in the manufacture of biofuel is to adapt bacteria to use a variety of substrates. In addition to these methods, the production of bioenergy from organic biomass and wastewater has gained traction thanks to microbial fuel cells and bioelectrical cells (Dai et al. 2016; Logan et al. 2015).

1.3.3   Treatment of Soil

The nitrogen cycle is crucial for converting atmospheric nitrogen into soil because it is facilitated by a variety of diazotrophs (Davison 1988; Basu et al. 2021; Sundari et al. 2019; Furukawa 2023). Symbiotic bacteria like *Mesorhizobium, Rhizobium, Sinorhizobium,* and *Bradyrhizobium,* found in leguminous roots, are among the key contributors to this process. Microbes not only aid in nutrient absorption and mineral availability for plants but also produce metabolites that stimulate plant growth and help in stress reduction (Hayat et al. 2010; Burd et al. 2000). Soil microbes are microscopic organisms that are vital for nutrient cycling, carbon sequestration, and maintaining soil health. They have the potential to mitigate climate change effects through carbon fixation and storage in organic matter. Understanding and managing these microbial communities can lead to improved soil health, water-holding capacity, and enhanced plant growth. Furthermore, it is important to note that soil microbes are instrumental in the process of breaking down substances, dealing with heavy metal contamination, and utilizing plants to restore polluted soils. However, climate change events can impact the diversity, composition, and metabolism of these microbes. The scientific community must acknowledge the significant role of soil microbes as crucial partners in addressing climate change. This can be achieved by adopting data-driven approaches to land management and implementing strategies to utilize their potential for climate mitigation, sustainable agriculture, and ecosystem resilience.

1.3.4   Oil Spills Treatment

Over the world's oceans, oil spills are becoming more frequent and alarming. Usually, human mistake occurs when transporting crude oil or refined petroleum products across several nations, leading to these accidents. As a result, addressing the remediation of oil spills has become a critical focus for countries and environmental agencies globally due to the significant and growing impact on the environment and wildlife. Natural processes such as petroleum seeps, coastal facilities, gasoline freighters, and offshore production or extraction wells also contribute to oil leakage in addition to human-caused leaks (Adebayo and Obiekezie 2018).

Oil spills have unfortunately become a frequent occurrence in oceans worldwide. Human mistake frequently causes these mishaps when transporting refined petroleum products or crude oil across nations. Because of the increasing harm that oil spills are causing to the environment and wildlife, countries and environmental organizations worldwide are paying close attention to the cleanup of oil spills. It is important to keep in mind that oil spills can occur naturally through processes like petroleum seeps, in addition to incidents involving coastal infrastructure, petroleum freighters, and offshore production or extraction wells. During an oil spill, large amounts of liquid hydrocarbons are inadvertently released into the environment, resulting in long-term pollution that harms the local ecology.

Tanker mishaps alone are thought to have leaked 5.72 million tons of crude oil into the ocean between 1970 and 2015. Tanker and pipeline accidents resulted in the loss of almost 7000 tons to the environment in 2015 (Kapagunta 2017). Though the most well-known, the 11 million gallons of crude oil that spilled during the Exxon Valdez event are not the largest oil spill in modern human history. During the Gulf War in 1991, Iraqi soldiers deliberately leaked between 380 and 250 million gallons of oil into the Persian Gulf, revealing the enormous environmental damage.

Oil spills have a significant effect on microbial communities (Dave and Ghaly 2011; Silva et al. 2022). When there is an oil leak and a layer of oil floats on the water's surface, a lot of marine animals die from poisoning and asphyxia, including fish, sea algae, marine mammals, and birds. Moreover, oil droplets that are dispersed and sink to the ocean floor harm the benthic community. The surprising abundance of hydrocarbons, which the surrounding microbial populations use as carbon and energy source, is also observed to cause rapid modifications in those communities.

The microbial degradation of oil's hydrocarbons contributes significantly to environmental cleanup after oil spills, and it has also been utilized to artificially expedite cleanup after significant spill incidents. Due to their greater efficiency, bacteria are the primary degraders of oil hydrocarbons, even though fungi, yeasts, and bacteria can all break down a complex mixture of oil hydrocarbons.

The ecology and composition of the local marine microbial population change dramatically after an oil spill. In these cases, the dominant species in the community are bacteria that facilitate the biodegradation processes by using various hydrocarbon molecules as a source of energy. Many studies have examined the nature and speed of changes in the local microbial communities in the Gulf of Mexico after the Deepwater Horizon and Exxon Valdez oil spills. Numerous facets of genetic and metabolic diversity, chemical and geological processes, community connections, and the background of pollution have been revealed by these investigations.

Table 1.4 lists the many species of bacteria and archaea that have been identified through metagenomics, culture-based research, and chemical methods as being able to break down hydrocarbons. Notably, several species of bacteria are known to be necessary hydrocarbon degraders, including *Oleispira, Cycloclasticus, Oleiphilus, Alcanivorax,* and *Thalassolituus.* Furthermore, it has been demonstrated that other families, like *Marinobacter* and *Pseudomonas*, have varying capacities for hydrocarbon breakdown.

Different types of microorganisms, such as bacteria, algae, fungi, and protozoa, live in different treatment systems. The kind of microbe that survives in garbage depends on the waste's properties. Table 1.5 provides a summary of several microorganisms and their functions in waste treatment.

Table 1.4 Oil-degrading bacteria

Acinetobacter
Afipia
Alcanivorax
Bacillus
Halanaerobium
Halomonas
Marinobacter
Microbulbifer
Methylophaga
Martelella
Paracoccus
Pseudomonas
Roseobacter
Roseovarius
Shewanella
Sphingomonas
Stappia
Vibrio

Based on Bordenave et al. (2007), Kostka et al. (2011), Vila et al. (2010), Gerdes et al. (2005)

Table 1.5 Different microorganisms capable of bioremediation of specific compounds

Microorganisms	Pollutants degraded by microorganisms
Penicillium chrysogenum	Monocyclic aromatic hydro carbons, benzene, toluene, ethyl benzene and xylene, phenol compounds
P. alcaligenes, P. putida, P. veronii, Achromobacter, Flavobacterium, Acinetobacter	Petrol and diesel polycyclic aromatic hydrocarbons tolueneMonocyclic aromatic hydrocarbons, e.g., benzene and xylene
A. niger, A. fumigatus, F. solani, P. Funiculosum, Alcaligenes odorans, Bacillus subtilis, Corynebacterium propinquum, Pseudomonas aeruginosa	Hydrocarbons, Phenols
Cyanobacteria, green algae and diatoms and Bacillus licheniformis	Naphthalene
Pseudomonas aeruginosa, P. putida, Arthobacter sp. *and Bacillus* sp.	Diesel oil
Bacillus cereus	Diesel oil
Pycnoporus sanguineous, Phanerochaete chrysosporium, Trametes trogii	Industrial dyes
Exiguobacterium indicum, Exiguobacterium aurantiacums, Bacillus cereus and Acinetobacter baumani	Azo dyes effluents
Saccharomyces cerevisiae	Heavy metals, lead, mercuryand nickel
Aspergillus versicolor, A. fumigatus, Paecilomyces sp.	Cadmium
Trichoderma sp., *Microsporum* sp., *Cladosporium* sp. *Aerococcus* sp., *Rhodopseudomonas palustris*	Pb, Cr, Cd

Parmar et al. (2021). Reproduced with permission

1.4   Benefits of Bioremediation Compared to Traditional Methods

Utilization of biological organisms for removing or neutralizing environmental toxins through metabolic processes is a fascinating method known as "bioremediation." These "biological" organisms encompass a variety of microscopic life forms, including fungi, algae, and bacteria. By treating the situation, or "remediating" it, these organisms are vital to the maintenance of the environment (Adebayo and Obiekezie 2018; Abatenh et al. 2017; Ayilara and Babalola 2023; Ganguly et al. 2023; Naz et al. 2023; Husain et al. 2022; Mondal and Palit 2019; Bala et al. 2022; Pant et al. 2023; Mani et al. 2020; Sharma et al. 2023, Sharma 2021).

Microorganisms are found to be thriving in an astounding range of habitats within the Earth's biosphere. These habitats include water, soil, animals, plants, deep sea habitats, and even freezing ice. Their large numbers and ability to absorb a wide range of contaminants make them ideal candidates to serve as environmental protectors.

Bioremediation technologies have been widely adopted in recent years and are expanding exponentially. Microbiological techniques are safe for the environment, which makes them a highly reliable and successful method for cleaning up contaminated areas (Parmar et al. 2021). Researchers have made significant strides in developing various bioremediation techniques aimed at successfully repairing contaminated areas with a cost-effective and environmentally responsible method.

Whether utilizing indigenous or non-indigenous microorganisms, the key lies in addressing the challenges associated with biodegradation and bioremediation of pollutants in polluted environments (Verma and Jaiswal 2016).

Two of the many benefits of bioremediation over chemical and physical remediation methods are its low cost and environmental friendliness.

Bioremediation changes dangerous substances into less harmful or non-toxic versions by using naturally occurring microbes.

The primary technique of bioremediation involves lowering, detoxifying, breaking down, mineralizing, or converting more dangerous pollutants into less dangerous ones.

The type of pollutant—which can include anything from heavy metals, greenhouse gases, hydrocarbons, pesticides, agrochemicals, and chlorinated compounds—determines how effective the removal method is in removing it.

Overall, bioremediation is essential for the elimination, decomposition, immobilization, or detoxification of various chemical pollutants and dangerous chemicals that are present in the surrounding environment due to the all-encompassing action of bacteria. Research studies that concentrate on using efficient procedures to clean up and reduce pollution at polluted places have drawn more attention in recent years. Interest in bioremediation has grown significantly since it is environmentally benign and compatible with the environment. This technique uses naturally occurring biological agents, like bacteria and green plants, to eliminate hazardous materials from the surroundings and bring it back to a non-hazardous condition.

Notably, bioremediation using microorganisms has demonstrated effectiveness even at very low pollutant concentrations, where traditional physicochemical methods have proven ineffective.

Frequently employed in bioremediation, microorganisms such as algae, bacteria, and fungi, can function alone or in combination to exhibit natural activity. Under certain circumstances, this microbial action might result in the mineralization, transformation, or immobilization of contaminants. In contrast to other physicochemical techniques that incur greater upfront and ongoing expenditures, microbial detoxification has reduced operating costs and doesn't require extra chemicals or catalysts, which prevents the production of potentially dangerous byproducts.

Because microbe-mediated detoxification procedures rely only on the natural biochemical and metabolic pathways of bacteria, no extra changes are needed. Because of this, they have garnered significant popular acceptance. Furthermore, with detoxification, pollutants are totally removed rather than just moved from one area of the environment to another. Microbe-based bioremediation is said to be the most economical method because it requires the least amount of equipment and space to operate. The nutritional flexibility of microorganisms makes them a good choice for bioremediation, and their ability to use various toxins as an energy source for development means that characterizing them from contaminated locations is a simple operation.

The bioremediation process can be readily induced by adjusting variables including pH, temperature, oxygen availability, moisture content, and duration of contact with the substrate. Overall, because of its effectiveness, affordability, and simplicity of application, the employment of microorganisms for pollutant bioremediation is a very promising technique.

Applications for microbial detoxification include biofiltration, phytoremediation, bioadsorption, and biotransformation, the latter of which is becoming more popular since it may change toxins into less harmful forms without producing any new pollutants. Microbes are a great tool for preserving the original natural ecosystems and reducing pollution because of all these reasons (Coelho et al. 2015; Abatenh et al. 2017; Kensa 2011). Several environmental parameters, including as pH, nutrient availability, and climate, have a substantial impact on the efficacy of bioremediation.

Conditions that are not ideal can make this procedure less successful. Not every kind of pollution may be remedied by bioremediation. Alternative cleanup techniques can be needed for chemicals that are extremely hazardous or resistant. Contaminated items are physically removed or treated as part of traditional remediation techniques.

Typically, these techniques involve chemical treatment, soil washing, incineration, and excavation. Conventional techniques are appropriate for the immediate containment of hazardous pollution since they can produce results more quickly. These techniques provide you more control over the rehabilitation process and increase the degree of assurance you have about the result. Many pollutants, including nonbiological chemicals like heavy metals and radioactive elements, can be addressed by traditional methods (Văcar et al. 2021; Cecchi et al. 2021).

There is a considerable risk of secondary pollution and environmental disturbance associated with the mechanical extraction and transportation of contaminated materials. Due to the substantial usage of heavy machinery and the need to dispose of polluted materials, traditional methods are frequently more expensive. These techniques do not support ecological restoration, and they frequently require the landfill disposal of polluted materials, which exacerbates long-term environmental problems. Conventional remediation techniques sometimes overlook the potential long-term ecological advantages of bioremediation in favor of immediate containment (Tudi et al. 2021).

The optimal course of action between bioremediation and traditional remediation approaches should be determined case-by-case. The kind and degree of contamination, site-specific factors, and project goals should all be considered in this research. Bioremediation provides a viable and eco-friendly substitute for conventional methods in less severe circumstances; these methods may be more appropriate in emergency or very toxic scenarios.

Utilizing the advantages of both techniques in a hybrid remediation plan can also be a workable way to restore sites in a way that is more eco-friendly and efficient.

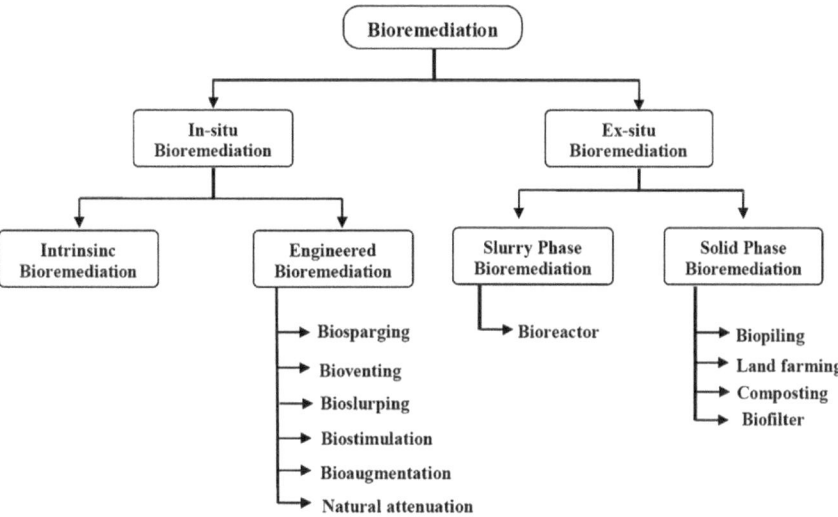

Fig. 1.5 Bioremediation approaches for environmental cleanup. Distributed under the terms of the Creative Commons Attribution License from Sharma (2021)

In the end, the decision should put ecosystem health and long-term environmental sustainability first. Many times, a well-rounded, efficient approach to site cleanup can be achieved by combining bioremediation with conventional remediation techniques. By minimizing their drawbacks, this hybrid approach maximizes the positive aspects of both approaches (Rosa et al. 2018).

The most dangerous materials can be quickly removed using conventional methods in situations where contamination is urgent or extremely poisonous.

This approach reduces environmental and public health risks while guaranteeing quick containment. To handle any residual contamination after the first cleanup, bioremediation techniques might be applied to the site. This facilitates the long-term microbial breakdown of contaminants, supporting the site's natural regeneration. Continuous monitoring is a common component of hybrid systems in order to evaluate the efficacy of both remedial techniques. It is possible to optimize the remediation process by making adjustments as needed. Hybrid tactics seek to balance quick containment with long-term sustainability by combining the best features of both bioremediation and conventional techniques, offering a more thorough and efficient site cleanup solution (Albright et al. 2017).

Bioremediation approaches for environmental cleanup are presented in Fig. 1.5 (Sharma 2021).

This book's main goal is to promote the use of diverse microbiological agents to treat polluted waste in a way that is both economical and environmentally sustainable, hence promoting environmental sustainability. It offers extensive information on environmental protection, industrial effluent treatment, and wastewater management.

References

Abatenh E, Gizaw B, Tsegaye Z, Wassie M (2017) The role of microorganisms in bioremediation—a review. Open J Environ Biol 2(1):030e046

Adebayo FO, Obiekezie SO (2018) Microorganisms in waste management. Res J Sci Technol 10(1):28–39

Albright VC 3rd, Wong CR, Hellmich RL, Coats JR (2017) Dissipation of double-stranded RNA in aquatic microcosms. Environ Toxicol Chem 36(5):1249–1253. https://doi.org/10.1002/etc.3648

Alexander M (1994) Biodegradation and bioremediation. Academic, San Diego

Angulo SC, Carrijo PM, Figueiredo AD, Chaves AP, John VM (2010) On the classification of mixed construction and demolition waste aggregate by porosity and its impact on the mechanical performance of concrete. Mater Struct 43(4):519–528

Ayilara MS, Babalola OO (2023) Bioremediation of environmental wastes: the role of microorganisms. Front Agron 5:1183691. https://doi.org/10.3389/fagro.2023.1183691

Bala S, Garg D, Thirumalesh BV, Sharma M, Sridhar K, Inbaraj BS, Tripathi M (2022) Recent strategies for bioremediation of emerging pollutants: a review for a green and sustainable environment. Toxics 10(8):484. https://doi.org/10.3390/toxics10080484

Banerjee A, Jhariya MK, Yadav DK, Raj A (2018) Micro-remediation of metals: a new frontier in bioremediation. In: Hussain C (ed) Handbook of environmental materials management. Springer, Berlin. ISBN: 978-3-319-58538-3. https://doi.org/10.1007/978-3-319-58538-3_10-1

Barea JM, Pozo MJ, Azcon R, Azcon-Aguilar C (2005) Microbial co-operation in the rhizosphere. J Exp Bot 56(417):1761–1778

Basu S, Kumar G, Chhabra S et al (2021) Role of soil microbes in biogeochemical cycle for enhancing soil fertility. In: Verma JP, Macdonald CA, Podille AR (eds) New and future developments in microbial biotechnology and bioengineering. Elsevier, Amsterdam, pp 149–157

Bordenave S, Goñi-Urriza MS, Caumette P, Duran R (2007) Effects of heavy fuel oil on the bacterial community structure of a pristine microbial mat. Appl Environ Microbiol 73(19):6089–6097

Brim H, McFarlan SC, Fredrickson JK, Daly JM, Venkateswaran A, Kostandarithes MH (2000) Engineering *Deinococcus radiodurans* for metal remediation in radioactive mixed waste environments. Nat Biotechnol 18:85–90

Burd G, Dixon DG, Glick BR (2000) Plant growth promoting bacteria that decrease heavy metal toxicity in plants. Can J Microbiol 46:237–245

Cecchi G, Cutroneo L, Di Piazza S, Besio G et al (2021) Port sediments: problem or resource? A review concerning the treatment and decontamination of port sediments by fungi and bacteria. Microorganisms 9:1279

Chowdhary P, Raj A, Bharagava RN (2018) Environmental pollution and health hazards from distillery wastewater and treatment approaches to combat the environmental. Chemosphere 194:229–246

Coelho LM, Rezende HC, Coelho LM, Sousa PA, Melo DF, Coelho NM (2015) Bioremediation of polluted waters using microorganisms. Adv Bioremed Wastewater Pollut Soil 10:60770

Dai H, Yang H, Liu X, Jian X, Liang Z (2016) Electrochemical evaluation of nano-$Mg(OH)_2$/graphene as a catalyst for hydrogen evolution in microbial electrolysis cell. Fuel 174:251–256

Daniel AJ, Raimondo Enzo E, Saez Juliana M, Costa-Gutierrez Stefanie B, Analía A, Benimeli Claudia S, Polti Marta A (2022) The current approach to soil remediation: a review of physicochemical and biological technologies, and the potential of their strategic combination. J Environ Chem Eng 10(2):107141

Dave D, Ghaly AE (2011) Remediation technologies for marine oil spills: a critical review and comparative analysis. Am J Environ Sci 7:424–440. https://doi.org/10.3844/ajessp.2011.424.440

Davison J (1988) Plant beneficial bacteria. Biotechnology 6:282–286

Demirbas A (2011) Waste management, waste resource facilities and waste conversion processes. Energy Convers Manag 52(2):1280–1287

Dunlop MJ (2011) Engineering microbes for tolerance to next-generation biofuels. Biotechnol Biofuels 4(32). https://www.biotechnologyforbiofuels.com/content/4/1/32

Furukawa K (2023) 'Super bugs' for bioremediation. Trends Biotechnol 21(5):187–190

Ganguly D, Prasanna KLV, Neelapu S, Goswami G (2023) Role of microbes in bioremediation. In: Verma P (eds) Industrial microbiology and biotechnology. Springer, Singapore. https://doi.org/10.1007/978-981-99-2816-3_19

Gerdes B, Brinkmeyer R, Dieckmann G, Helmke E (2005) Influence of crude oil on changes of bacterial communities in Arctic sea-ice. FEMS Microbiol Ecol 53(1):129–139

Giusti L (2009) A review of waste management practices and their impact on human health. Waste Manage 29:2227–2239

Gu JD, Ford TE, Mitton DB, Mitchell R (2000) Microbial degradation and deterioration of polymeric materials. In: Revie W (ed) The Uhlig corrosion handbook, 2nd edn. Wiley, New York, pp 439–460

Hassan BA, Venkateshwaran AA, Fredrickson JK, Daly MJ (2003) Engineering *Deinococcus geothermalis* for bioremediation of high temperature radioactive waste environments. Appl Environ Microbiol 69:4575–4582

Hayat R, Ali S, Amara U et al (2010) Soil beneficial bacteria and their role in plant growth promotion: a review. Ann Microbiol 60:579–598. https://doi.org/10.1007/s13213-010-0117-1

Holt MS (2001) Sources of chemical contaminants and routes into the freshwater environment. Food Chem Toxicol 38:S21–S27

Husain R, Vikram N, Yadav G, Kumar D, Pandey S, Patel M, Khan NA, Hussain T (2022) Microorganisms: an eco-friendly tools for the waste management and environmental safety. In: Arora S, Kumar A, Ogita S, Yau YY (eds) Biotechnological innovations for environmental bioremediation. Springer, Singapore. https://doi.org/10.1007/978-981-16-9001-3_36

Hussain T, Dhanker R (2021) Science of microorganisms for the restoration of polluted sites for safe and healthy environment. In: Shah M, Rodriguez-Couto S (eds) Microbial ecology of wastewater treatment plants. Elsevier, Amsterdam, pp 127–144. https://doi.org/10.1016/C2019-0-04695-X

Inslee J, Hendricks B (2009) Apollo's fire: igniting America's clean energy economy. Island Press. 86094-332-2

Jhariya MK, Yadav DK, Banerjee A (2018) Plant mediated transformation and habitat restoration: phytoremediation an eco-friendly approach. In: Gautam A, Pathak C (eds) Metallic contamination and its toxicity. Daya Publishing House, A Division of Astral International Pvt. Ltd, New Delhi, pp 231–247. ISBN: 9789351248880

Kapagunta C (2017) Role of microbes in oil spills remediation and degradation of hydrocarbons. https://www.projectguru.in/role-microbes-oil-spills-remediation-degradation-hydrocarbons/

Kensa VM (2011) Bioremediation-an overview. J Ind Control Pollut 27(2):161–168

Kostka JE, Prakash O, Overholt WA, Green SJ, Freyer G, Canion A, Delgardio J, Norton N, Hazen TC, Huettel M (2011) Hydrocarbon-degrading bacteria and the bacterial community response in gulf of Mexico beach sands impacted by the deepwater horizon oil spill. Appl Environ Microbiol 77(22):7962–7974

Logan BE, Wallack MJ, Kim KY et al (2015) Assessment of microbial fuel cell configurations and power densities. Environ Sci Technol Lett 2:206–214. https://doi.org/10.1021/acs.estlett.5b00180

Maghraby DM, Hassan J (2018) Heavy metals bioaccumulation by the green alga *Cladophora herpestica* in Lake Mariut, Alexandria. Egypt J Pollut 1:1

Mani S, Chowdhary P, Zainith S (2020) Microbes mediated approaches for environmental waste management. In: Chowdhary P, Raj A, Verma D, Akhter Y (eds) Microorganisms for sustainable environment and health. Elsevier, pp 17–36

Martinez AG, Sihvonen M, Palazon BM, Sanchez AR, Mikola A, Vahala R (2018) Microbial ecology of full-scale wastewater treatment systems in the polar Arctic circle: Archaea, Bacteria and Fungi. Sci Rep 8:2208. https://doi.org/10.1038/s41598-018-20633-5

Matuszewska A (2016) Microorganisms as direct and indirect sources of alternative fuels. https://www.intechopen.com/books/alternative-fuels-technical-and-environmental-
conditions/microorganisms-as-direct-and-indirect-sources-of-alternative-fuels

Meng X, Yang J, Xu X, Zhang L, Xian M, Nie Q (2009) Biodiesel production from oleaginous
microorganisms. Renew Energy 34(1):1–5

Mondal S, Palit D (2019) Effective role of microorganism in waste management and environmental
sustainability. In: Jhariya M, Banerjee A, Meena R, Yadav D (eds) Sustainable agriculture,
forest and environmental management. Springer, Singapore. https://doi.org/10.1007/978-981-
13-6830-1_14

Moradian JM, Fang Z, Yong YC (2021) Recent advances on biomass-fueled microbial fuel cell.
Bioresour Bioprocess 8:14. https://doi.org/10.1186/s40643-021-00365-7

Nandan A, Yada BP, Baksi S, Bose D (2017) Recent scenario of solid waste management in India
66:56–74

Narwal E, Choudhary J, Singh S, Nain L, Kumar S, Dotaniya ML, Kumar A (2020) Microbial
biofuels: renewable source of energy. Front Soil Environ Microbiol 19:181–192

Naz M, Raza MA, Zaib S, Tariq M, Afzal MR, Hussain S, Dai Z, Du D (2023) Major groups of
microorganisms employed in bioremediation. In: Bhat RA, Butnariu M, Dar GH, Hakeem KR
(eds) Microbial bioremediation. Springer, Cham. https://doi.org/10.1007/978-3-031-18017-0_
8

Ochedi FO, Liu Y, Adewuyi YG (2020) State-of-the-art review on capture of CO_2 using adsorbents
prepared from waste materials. Process Saf Environ Protect 139:1–25

Pant R, Gupta A, Singh A, Srivastava S, Patrick N (2023) A summary of the role of microorganisms in
waste management. In: Debbarma P, Kumar S, Suyal DC, Soni R (eds) Microbial technology for
sustainable e-waste management. Springer, Cham. https://doi.org/10.1007/978-3-031-25678-3_
21

Parmar S, Daki S, Bhattacharya S, Shrivastav A (2021) Microorganism: an ecofriendly tool for
waste management and environmental safety. In: Shah MP, Rodriguez-Couto S, Kapoor RT
(eds) Development in wastewater treatment research and processes, chap. 8. Elsevier 2022, pp
175–193. ISBN: 9780323856577. https://doi.org/10.1016/B978-0-323-85657-7.00001-8.

Raj A, Jhariya MK, Bargali SS (2018a) Climate smart agriculture and carbon sequestration. In:
Pandey CB, Gaur MK, Goyal RK (eds) Climate change and agroforestry: adaptation mitigation
and livelihood security. New India Publishing Agency (NIPA), New Delhi, pp 1–19. ISBN:
9789-386546067

Raj A, Jhariya MK, Harne SS (2018b) Threats to biodiversity and conservation strategies. In: Sood
KK, Mahajan V (eds) Forests, climate change and biodiversity, vol 381. Kalyani Publisher, New
Delhi, India, pp 304–320

Rosa C, Kuo YW, Wuriyanghan H, Falk BW (2018) RNA interference mechanisms and applications
in plant pathology. Annu Rev Phytopathol 56:581–610. https://doi.org/10.1146/annurev-phyto-
080417-050044

Satyanarayana T, Johri BN, Prakash A (eds) (2012) Microorganisms in environmental management:
microbes and environment. Springer Science & Business Media

Saxena N, Bhargava R (2017) A review on air pollution, polluting agents and its possible effects in
21st century. Adv Bio Res 8(2)

Sharma M, Agarwal S, Agarwal Malik R, Kumar G, Pal DB, Mandal M, Sarkar A, Bantun F,
Haque S, Singh P, Srivastava N, Gupta VK (2023) Recent advances in microbial engineering
approaches for wastewater treatment: a review. Bioengineered 14(1):2184518. https://doi.org/
10.1080/21655979.2023.2184518

Sharma S, Shah KW (2005) Generation and disposal of solid waste in Hoshangabad. In: Proceedings
of the 2nd international congress of chemistry and environment, Indore, India, pp 749–751

Sharma I (2021) Bioremediation techniques for polluted environment: concept, advantages,
limitations, and prospects. IntechOpen. https://doi.org/10.5772/intechopen.90453

Silva IA, Almeida FCG, Souza TC et al (2022) Oil spills: impacts and perspectives of treatment technologies with focus on the use of green surfactants. Environ Monit Assess 194:143. https://doi.org/10.1007/s10661-022-09813-z

Singh Pandya R, Kaur T, Bhattacharya R, Bose D, Saraf D (2023) Harnessing microorganisms for bioenergy with microbial fuel cells: powering the future. Water-Energy Nexus 7:1–12

Sundari SK, Prakash A, Yadav P et al (2019) Plant growth-promoting microbes as front-runners for on-site remediation of organophosphate pesticide residues in agriculture soils. In: Arora NK, Kumar N (eds) Phyto and rhizo remediation. Springer, Singapore, pp 249–285

Tudi M, Daniel Ruan H, Wang L, Lyu J, Sadler R, Connell D, Chu C, Phung DT (2021) Agriculture development, pesticide application and its impact on the environment. Int J Environ Res Public Health 18(3):1112. https://doi.org/10.3390/ijerph18031112

Văcar CL, Covaci E, Chakraborty S, Li B et al (2021) Heavy metal-resistant filamentous fungi as potential mercury bioremediators. J Fungus 7:386

Varma D, Meena RS, Kumar S, Kumar E (2017) Response of mung bean to NPK and lime under the conditions of Vindhyan region of Uttar Pradesh. Leg Res 40(3):542–545

Venkatesh S, Krishnaveni M (2020) Microbes: the next-generation bioenergy producers. In: Kashyap BK, Solanki MK, Kamboj DV, Pandey AK (eds) Waste to energy: prospects and applications. Springer, Singapore. https://doi.org/10.1007/978-981-33-4347-4_2

Verma JP, Jaiswal DK (2016) Book review: advances in biodegradation and bioremediation of industrial waste. Front Microbiol 6:1–2. https://doi.org/10.3389/fmicb.2015.01555

Vila J, María Nieto J, Mertens J, Springael D, Grifoll M (2010) Microbial community structure of a heavy fuel oil-degrading marine consortium: linking microbial dynamics with polycyclic aromatic hydrocarbon utilization. FEMS Microbiol Ecol 73(2):349–362

Weltens R, Vanermen G, Tirez K, Robbens J, Deprez K, Michiels L (2012) Screening tests for hazard classification of complex waste materials—selection of methods. Waste Manag 32(12):2208–2217

Wu Q, Jiao S, Ma M et al (2020) Microbial fuel cell system: a promising technology for pollutant removal and environmental remediation. Environ Sci Pollut Res 27(7):6749–6764

Yadav MD, Patil SV, Shinde PP (2023) Solid wastes: types, sources, management. IJCRT 11(4):15–23

Chapter 2
Strategies and Technologies for Waste Management

Abstract Bioremediation is an eco-friendly process that uses natural and genetically modified microorganisms to break down hazardous waste. It can be done on-site or off-site, using pure cultures or mixed consortia. Microbial phytoremediation is a cost-effective and environmentally cleaner alternative to landfill burning. Microbial consortia offer several benefits in water treatment. Strategies for managing microbial waste and emerging innovations in microbial consortium technology are discussed.

Keywords Bioremediation · Bioleaching · Bioaugmentation · Biostimulation · Bioventing · Biopiles · Biofiltration · Phytoremediation · Microbial consortium · Microalgae · Bacteria · Wastewater treatment · Algal-bacterial consortium · Green technology

2.1 Various Strategies for Managing Microbial Waste

It is necessary to handle the surplus trash produced by human activity in a safe and sustainable way. Rather than relying solely on physicochemical methods, we are transitioning toward eco-friendly and efficient processes such as bioremediation (Kadri et al. 2017). It encompasses various approaches that utilize natural and genetically engineered microbes for breaking down hazardous waste, with or without the presence of oxygen (Parmar et al. 2021; Ganguly et al. 2023; Husain et al. 2022). These procedures may be completed either at the location or off it using pure cultures or mixed consortia of microorganisms. Because of their extraordinary metabolic capacities, microbes are essential to the breakdown of toxic substances because they can use them as fuel for development through co-metabolism, fermentation, and both aerobic and anaerobic respiration. Table 2.1 lists the main bioremediation techniques (Oyetibo et al. 2017), while subsequent sections elaborate on the major bioremediation approaches.

Table 2.1 Categorization of various bioremediation techniques according to their operational sites

Bioremediation								
In situ					Ex situ			
Natural	Engineered				Engineered			
Bioremediation	Biosparging/ Bioslurping	Bioaugmentation	Biological reactive barriers	Anaerobic reductive dechlorination	Composting	Constructed wetlands	Slurry/ aqueous bioreactors	Biopiles

Based on Oyetibo et al. (2017), Parmar et al. (2021)

2.1.1 Bioleaching

Metals are a significant cause of pollution and make up a sizable portion of the environmental pollution load. Many physiochemical technologies have been created for the removal of metals, but due to their high cost of operation and maintenance, their popularity has declined. In contrast, bioleaching speeds up the process of removing metals from liquids by utilizing the metabolic power of microorganisms. Bioleaching requires microbes that can survive in harsh circumstances with high metal ion concentrations, a strongly acidic pH, and oxidizing conditions.

Microbes oxidize mineral sulfides in this process to produce the required energy. In order to successfully remove metals including cadmium, nickel, cobalt, lead, copper, iron, and anitimony, bioleaching can oxidize them either directly or indirectly. *Thiobacillus thioxoxidans, Acidithiobacillus ferrooxidans,* and *T. cuprinicus* are important bacterial cultures in bioleaching that help oxidize metal sulfides (Kelly and Wood 2000).

Fungi have two possible roles in the bioleaching process. This can be achieved by a mycorrhizal relationship or the white-rot fungus Saprophyta. For fungal detoxification, genera such as *Trichoderma, Fusarium, Penicillium* and *Sctyalidium* are utilized (Barclay et al. 1998). Plant roots' mycorrhizal relationships with fungi can be utilized as a bioleaching marker and a sign of contaminated soil. Microbes that have been isolated from various ecosystems and possess particular traits are regarded as exceptional bioagents because of their resistance to metal stress. Ecological cycle maintenance and environmental purification can both be greatly aided by bioleaching.

2.1.2 Bioaugmentation

A viable long-term remedy for treating pollution is bioaugmentation. This strategy includes adding efficient and targeted microbial strains into the pre-existing microbial community that can degrade contaminants. Improving the capacity of the entire microbial consortia to biodegrade pollutants is the aim. When extremely complex molecules that are resistant to biodegradation are present in the system, bioaugmentation is required since they can lower the process's efficiency and quality. Bioaugmentation offers an affordable and eco-friendly substitute for physiochemical processes by circumventing the drawbacks of biodegradation (Mrozik and Piotrowska-Seget 2010; Tyagi et al. 2011). There are situations when a number of elements combine to make pollutants resistant to biodegradation.

A few examples of these characteristics are the ineffectiveness of microbial enzymes as substrates, increased stability, decreased water solubility, decreased bioavailability, increased toxicity, chemical configurations, and the creation of novel compounds. In these circumstances, bioaugmentation could be a helpful tactic since it allows treatment to be targeted specifically at a contaminant that is typical in a specific setting. Common genera of bacteria employed in bioaugmentation include *Desulfobacterium, Acinetobacter, Burkholderia, Pseudomonas, Serratia, Aspergillus,* and *Sphingomonas.* Remarkably, the efficacy of bioaugmentation can be enhanced by the use of nanotechnology, genetically altering bacteria, quorum sensing, and immobilized cells (Nzila et al. 2016).

2.1.3   Enhancing Biological Activity

One remediation strategy that is reasonably priced is biostimulation. By introducing rate-limiting nutrients to heavily contaminated areas, this strategy aims to increase the activity of pre-existing microbes and enhance their capacity to degrade hazardous pollutants. By giving the microorganisms these nutrients, the decontamination process is facilitated and the microbes' metabolism is directly affected, which raises the breakdown rate. This method is frequently applied to locations contaminated by petroleum, as the microbial community there has low metabolism and needs more nutrients to function more actively. Organic matter rich in nutrients can be added to accelerate degradation in addition to the essential nutrients that regulate the pace of degradation.

Many pollutants, including polymers like polyesters and polyurethanes, sulfate compounds, and petroleum compounds can be effectively removed through biostimulation. Animal excrement, sewage sludges, leftover vegetation, ground-up seeds, composted cellulose or straw, cornmeal, chicken dung, and dairy manure, and substances like glucose, potassium, phosphate, nitrate, and ammonia salts are among the nutrients that are frequently utilized for biostimulation. Crucially, because biostimulation makes use of the native bacteria already present in the environment, it doesn't harm it. It is critical to carefully evaluate the properties of native bacteria as well as ambient variables like pH, humidity, and temperature before beginning biostimulation. Furthermore, the effectiveness of biostimulation can differ depending on the surrounding environment; for example, bacteria in marine habitats degrade less efficiently than those on soil. Thus, under these situations, biostimulation turns out to be a useful tactic (Goswami et al. 2018).

2.1.4   Bioventing

One popular bioremediation technique is called "bioventing," which is opening up the airflow to provide the bioremediation system additional oxygen. Achieving an adequate oxygen supply through airflow control can greatly increase microbial activity. It's critical to regulate the rate at which air is introduced into the system.

It is advised to keep the airflow rate gradual and constant to avoid the system becoming saturated too soon. Additionally, sometimes modifications are made to the system through the addition of nutrients, known as biostimulation. The success of bioventing largely depends on efficiently dispersing and redistributing air over the surface in order to address high degradation rates. This is a critical aspect that must be well managed for optimal results. This technique is widely used for in situ remediation because it can enhance the biodegradation rate of microbes by providing necessary aeration. Achieving desired results in bioventing requires injecting air from multiple points and ensuring uniform distribution. Depending on the refractory molecules in the system, bioventing may also entail low amounts of nitrogen, carbon dioxide, and hydrogen. It is not always necessary to provide oxygen alone. Instead of a saturated system, aeration is continuously supplied to the unsaturated zone. Bioventing is widely utilized in the cleanup of hydrocarbon-contaminated sites, chlorine compound-contaminated sites, and locations where diesel has been removed (Azubuike et al. 2016).

2.1.5 Biopiles

Biopiling is a highly effective method for treating various petrochemical-contaminated soils outside of their original location. This technique involves gathering contaminated soils from different areas and forming them into piles at a single location. The pace at which the bacteria biodegrade in the piles is greatly accelerated by establishing ideal conditions for microbial development. To increase biopiling's overall efficacy, other strategies including biostimulation, bioaugmentation, and phytoremediation can be used. Aerobic bacteria predominate among those engaged in the biodegradation process. These biopiles are made up of mounds of dirt and polluted or dried sediments that were gathered from different locations where garbage was produced. It is crucial to give nutrients like nitrates and phosphates, maintain pH levels, guarantee adequate moisture, and introduce oxygen to promote microbial biodegradation.

By adding bacteria that break down hydrocarbons into the system, biopiling advances by directly stimulating the piles. Reduced cleanup costs following biopile building, low monitoring expenses, and very simple long-term restoration of ecological processes at the site are some benefits of biopiling. Large-scale hydrocarbon-impacted soil remediation has become popular when ecopiling is paired with phytoremediation and passive biopiling technologies. According to Germaine et al. (2015), there is a good chance that contaminated areas can be successfully remedied using this combination strategy.

2.1.6 Biofiltration

As part of the biofiltration setup, microorganisms are cultivated on porous materials such as soil peat, compost, or a mix of these. There is a thin coating of water, called "biofilm," surrounding the filter media and microbial culture. Odors, volatile organic compounds (VOCs), and air pollutants can all be effectively managed by biofiltration, which is why it is gaining popularity as a pollution control technique. It can be used to purify drinking water since it not only removes tiny particles but also uses microbiological action to break down dissolved organic materials. Microbiological and physical processes are combined in biofiltration, which mainly relies on microbial decomposition.

When employing a biofilter, certain process fundamentals must be followed in addition to controlling particular parameters for best outcomes. Prior to designing a biofilter, it is essential to comprehend its process engineering (Wani et al. 1997). In a biofiltration system, the waste to be treated is deposited onto a fixed bed. Plastic beads, compost, dirt, peat, or any other equally solid matrix that supports microorganisms can be used to create this bed.

Enough nutrients are added with this setting. The bacteria subsequently oxidize the pollutants in the waste, producing mineral salts, CO_2, and water as a result.

Some ideas involve immobilizing the degrading microorganisms on inert and porous carriers, like calcium alginate and polypropylene pellets, as technology progresses and the procedure is perfected. On a surface this permeable, immobilizing the bacteria can increase their survivability. The trash that is fed into the system provides the heterotrophic bacteria in the filter bed with energy and carbon for their microbial metabolism. Ammonia and hydrogen sulfide are two inorganic substances that autotrophic bacteria can handle with ease. Because there is a lot of surface area accessible for swapping out components, transformation occurs during the process, necessitating the supply of oxygen as needed. When transformation is finished, byproducts like acid metabolites or mineral salts are created, depending on the pollutant handled (Leson and Winer 1991).

2.1.7   Utilizing Microorganisms to Aid Phytoremediation

Using soil microbes and plants, phytoremediation is an environmentally sustainable approach that removes both organic and inorganic toxins from a variety of settings. It encourages the biodegradation of pollutants in addition to facilitating their absorption, accumulation, and immobilization. Phytoremediation is a holistic technique to cleaning up polluted places since it changes the chemical, physical, and biological characteristics of the soil while promoting the native microbial community. There are various choices involved in phytoremediation (Table 2.2). Metallophytes are often employed in phytoremediation, especially when dealing with metal pollution. In order to improve remediation efficiency, hyperaccumulating plants are strategically combined with particular bacteria in a process known as microbe-assisted phytoremediation, particularly when the plants are under stress. This is accomplished by supplying the soil with the essential nutrients and plant growth-promoting rhizobacteria (PGPRs) to promote the remediation potential of the plants.

Another bioremediation method is composting, which involves introducing organic materials to contaminated soil in order to promote the development of microbial colonies. Similar to this, land-farming is a technique in which pollutants are

Table 2.2 Various types of phytoremediation techniques

Phytostabilization
Plant roots play a crucial role in restricting the spread and uptake of pollutants in the soil
Phytovolatilization
Pollutant transformation into a volatile form and subsequent release into the atmosphere
Phytoextraction
Pollutant buildup in biomass that can be harvested, such as shoots
Phytofiltration
Pollutant sequestration by plants from contaminated waters
Phytodegradation
Organic xenobiotics are broken down in plant tissues by plant enzymes
Rhizodegradation
Degradation of organic xenobiotics by rhizosphere microorganisms
Phytodesalination
Removal of surplus salts from saline soils by halophytes

Based on Ali et al. (2013), Parmar et al. (2021)

incorporated into the soil and converted into non-toxic forms by natural microbes with the help of degradative enzymes. With so many different bioremediation techniques available, it can be difficult to choose the best strategy without a thorough grasp of the benefits and drawbacks of each technique. To select the best bioremediation technique for a given set of site conditions, comprehensive knowledge about the overall features and benefits of several techniques is crucial.

Criteria for selecting bioremediation techniques, including their respective benefits and drawbacks are presented below (Oyetibo et al. 2017):

2.1.7.1 Oxidative Biodegradation in situ

Transfer of contaminants from the sediment-sorbed phase to the aqueous phase, followed by the enzymatic biodegradation of these contaminants within the sediment-sorbed phase. This process is influenced by site-specific conditions and the availability of active biopolymers.

Characterized as the most cost-effective, non-invasive, and relatively passive method, relying on natural attenuation processes.

Faces challenges such as environmental constraints, prolonged treatment durations, and difficulties in monitoring.

2.1.7.2 Phytoremediation

Geological and climatic features of the treatment site.

Relatively inexpensive, straightforward to implement and maintain, and employs multiple mechanisms for contaminant removal. There is no requirement for disposal sites, and it is visually appealing as well as environmentally sustainable.

Contaminants that have accumulated may be reintroduced into the environment during litter fall, and the process may take longer compared to alternative technologies. Additionally, there is a potential for accumulation in fuel woods, which may enhance the solubility of the contaminants.

2.1.7.3   Biostimulation: Biosparging, Bioventing

The theoretical foundation for increasing nutrient production is established by the site conditions in terms of physicochemical elements, ecological stoichiometry (supply rates and ratios of nutrients in relation to the nutritional requirements of the inherent physiology of the cell), diversity of native microorganisms, and ecophysiological state of pollutant-degrading microorganisms degrading pollutants.

The equipment's accessibility and simplicity of installation provide quick treatment times and seamless integration with other technologies.

High contamination levels can be harmful to living beings, while low remediation limits are not usually achievable.

2.1.7.4   Bioaugmentation Biodetoxification

Native microorganisms' capacity for biodegradation; the presence of antibiosis-inducing factors; the biodegradability of persistent organic pollutants (POPs); chemical solubility; the dispersal of contaminants; exposure history and the adaptability of microbial degraders.

The most affordable, non-invasive, passive natural attenuation methods available.

Environmental limitations, an extended term of therapy, challenges with monitoring, and the potential for a prolonged lag period.

2.1.7.5   Phytoextraction Rhizofiltration

Similar to phytoremediation, biostimulation

Characterized by low expenses.

Have certain limitations, including the need for ample space, prolonged treatment durations, the necessity to manage abiotic losses, challenges related to mass transfer, and constraints concerning bioavailability.

2.1.7.6 Slurry Reactors and Aqueous Reactors

Must consider the toxicity of modifications, hazardous concentrations of pollutants, and bioaugmentation variables.

Characterized by constant performance, quick kinetics of degradation, and ideal environmental conditions that enhance mass transfer and make it easier to apply inoculants and biopolymers.

Because of the necessary excavation required by the associated matrices (water and sediments), the capital and operating costs are rather significant.

2.2 Emerging Innovations in Microbial Consortia Technology

Worldwide, water contamination from diverse sources has grown to be a major issue. An effective substitute for treating water contamination is the employment of microorganisms. Currently in use as an innovative and evolutionary wastewater treatment method is microbial remediation. Microorganisms, encompassing fungi, bacteria, yeasts, and algae are naturally present in the environment and provide a sustainable approach to addressing the issue of water pollution.

Possible limitations on the application of microorganisms during the treatment procedure include a lack of knowledge about the metabolic capacity of microbes to decompose pollutants and an absence of regulated environments such as pH, temperature, the proper quantity of contaminants, essential elements, and extended duration of use. If the process is not handled, contaminants that are not completely broken down in the process may produce toxic consequences. Therefore, precise inside characterization is one possible way to address the issues with microbial-assisted wastewater treatment. Genetic engineering advancements have also contributed to the success of microbial wastewater treatment. Microbial strains that are engineered to possess strong metabolic functions and clearly defined detoxification pathways will certainly play a significant role in mitigating the challenges posed by wastewater contamination.

When breaking down complex compounds, microbial consortia are preferable to isolated bacteria because they can provide each enzyme needed for the biodegradation pathway with an appropriate catalytic environment and are more stable and adaptable in the developing environment. Advances in synthetic biology and gene editing technologies have made it possible to create artificial microbial consortia systems that exhibit greater stability, resilience, and efficacy. Their significant degradation capacity also makes them useful for the production of high-value chemicals. Additionally, it has been demonstrated that microbial consortium systems have promise for the breakdown of complicated substances.

2.2.1   Microbial Consortium-Based Technology

The new green strategy based on biotechnology is called the *Microbial Consortium*. When treating wastewater with a single microbe strain, efficiency may suffer and results may not be as good. As a result, the application of microbial consortia has been suggested by numerous research findings (Hosseinzadeh et al. 2020; Ji et al. 2020). A good option may be to form consortiums with various populations of environmental microorganisms that are able to break down contaminants in wastewater. When compared to applying a single strain, these consortiums offer numerous benefits, including quick removal, help with treated wastewater secondary application, and the promotion of ecological sustainability.

The volume of sewage released by different companies worldwide is becoming a major environmental concern due to the globalization of the economy and contemporary industrialization (Sierra et al. 2018; Kong et al. 2018). Conventional treatment systems are typically costly, require enormous energy consumption, and frequently still fail to address all sewage-related issues. A reasonably clean and effective method of treating sewage is the use of microbial consortia, particularly when treating eutrophication in the sewage system (Plöhn et al. 2021).

Complex chemicals such as petroleum, plastics, azo dyes, antibiotics, and some contaminants found in sewage can be broken down by microbial consortia (Fig. 2.1). Biofilm is created in the natural environment when several microbial groups come together and are linked by exopolymeric materials. The system functions as a whole, with microbial partners helping to build a robust community. One new method for treating wastewater is the formation of consortiums. Because of its capacity to refine biomass and low power consumption, the algal–bacterial consortium offers numerous benefits (Kang et al. 2017).

The microbial community's basic idea is to use mutually beneficial connections to encourage the elimination of pollutants from wastewater. According to Kujan et al. (2006), there is a synergistic link between bacteria and algae in the removal of BOD and nitrogen and phosphorus by absorption. A good foundation for bioremediation is provided by the link that has been formed between bacteria and algae (Lee et al. 2016). Cyanobacteria bacteria carry out photosynthesis, which changes the inorganic carbon in wastewater into organic carbon (Li et al. 2007). Photosynthetic algae use CO_2 generated by bacterial oxidation as their carbon supply. Algae development is aided by decomposers such as *Acinetobacter*, which may oxidize organic carbon sources and eliminate BOD (Lim et al. 2010). Numerous studies' conclusions have backed up the idea of using microbial consortiums to treat wastewater (Liu and Chen 2014). The biological elimination of nitrogen-containing substances by bioaugmentation was made easier by the utilization of the *Ecobacter* bacterial consortium (Malik and Ahmed 2012). It was observed that ammonium ions underwent a transformation due to a reduction reaction mediated by microorganisms, which resulted in a diminished level of ammonium ions present at the conclusion of the treatment period. The optimization and application of a well-defined microbial consortium, comprising

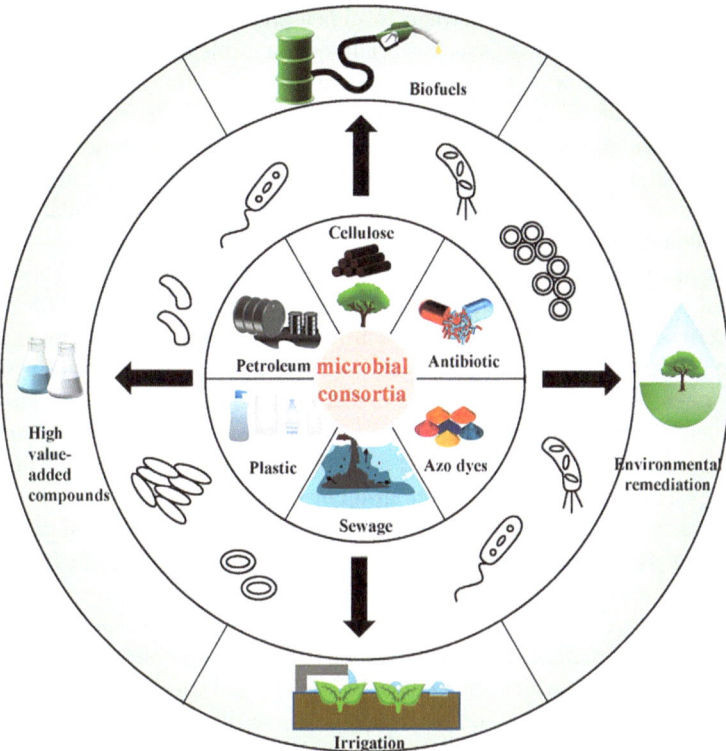

Fig. 2.1 Microbial consortia bioremediation and reuse of complex compounds Cao et al. (2022). Distributed under the terms of the Creative Commons Attribution License (CC BY)

both algae and bacteria, within the phycosphere has shown significant effectiveness in the advanced treatment of wastewater (Meckenstock 2016).

Microalgae and bacteria working together to treat paper pulp effluent resulted in a microbial consortium that removed nutrients and organic debris with high efficiency (Monica et al. 2016). The investigation by Rehman et al. (2018) focused on the application of a microbial consortium for the treatment of oil field wastewater, featuring *Acinetobacter junii* TYRH47, *Bacillus subtilis* LOR166, *Klebsiella* sp. LCR187, and *Acinetobacter* sp. BRS156. Using a microbial consortia, a removal efficiency of more than 90% of contaminants from textile effluent has been reported by Tara et al. (2019). Leong et al. (2019) indicated that the application of a microalgae consortium in conjunction with bacteria achieved a 94% efficiency in the elimination of pollutants from municipal wastewater. Microbes secrete different enzymes and organic acids to carry out the breakdown.

Effective Microorganism (EM), which includes *Saccharomyces, Aspergillus, Streptomyces, Lactobacillus*, and *Pseudomonas*, was employed by Monica et al. (2011) to biodegrade the sewage load in the water. In this consortium, cellulose and lignin are broken down by *Lactobacillus*. Bioactive chemicals released by *Pseudomonas* operate on sewage to either precipitate or detoxify the metal. *Aspergillus* breaks down organic materials quickly, releasing esters and alcohol. Table 2.3 illustrates how well microbial consortia are used to clean wastewater from different sources. Bacterial-algae consortia, which are derived from activated sludge, are the microbial consortia that are frequently researched for wastewater treatment. Microalgae have metabolic flexibility, enabling them to transition between autotrophic and heterotrophic metabolic modes in response to available carbon sources and nutrients in their local environment.

Consequently, a lot of researchers are interested in using microalgae to form microbial consortia in water (Kumar et al. 2010, 2021; Raja et al. 2008; Subash Chandra Bose et al. 2013; Li et al. 2016; Wijffels et al. 2010). In contrast to bottom-up approaches, top-down methods for the development of microbial consortia in wastewater treatment have demonstrated greater stability and reduced need for screening. Consequently, this topic will not be extensively discussed here. The combination of microalgae and bacteria has exhibited a beneficial stimulating effect on microbial consortia established through the bottom-up method.

Algal secretions serve as the primary supply of carbon (in the form of proteins, lipids, and carbs) for bacteria, while carbon dioxide and the oxygen in algae are advantageous for bacterial growth. Bacterial metabolites can be employed as growth stimulants for algae. Furthermore, bacteria may be able to find a stable home on the cell surface of microalgae (Ramanan et al. 2016). Microalgae require the growth of vitamin B12 and auxin which are secreted by bacteria and which also convert organic matter into mineral forms (Salim et al. 2014). As a result, the collaboration between microorganisms and microalgae leads to a more effective detoxification of organic and inorganic contaminants, as well as a superior removal of nutrients from sewage compared to their individual efforts (Wijffels et al. 2010; Subash Chandra Bose et al. 2013; Xiong et al. 2017).

In one study, the effects of varying amounts of phosphorus and arsenic on the transformation and accumulation of arsenic in pellets of *Aspergillus oryzae, Chlorella vulgaris*, and bacto-algae were experimentally compared. The bacto-algae ball had the best removal effectiveness of all the treatments and the strongest capacity to accumulate arsenic (Li et al. 2019). The eutrophication of wastewater has been shown to benefit from the use of constructed microbial communities.

Mujtaba et al. (2017) investigated the concurrent elimination of nutrients, specifically phosphate and ammonium, along with chemical oxygen demand (COD) within a co-culture system comprising *Chlorella vulgaris* and *Pseudomonas putida. P. putida's* ability to absorb nutrients was enhanced in the consortia, as evidenced by the fact that the co-culture system removed more nutrients and COD than each independent culture system. Therefore, there are several opportunities for using microbial consortia and microalgae together in sewage treatment.

Even though the application of microbial consortiums, such as microalgae-bacterial systems, for wastewater treatment has been the subject of numerous studies, more investigation is still required to optimize parameters for large-scale units. The primary difficult issue is keeping the consortium stable. Certain factors, such as

Table 2.3 Microbial consortium for the treatment of wastewater

Consortium species	Genus	Complex wastewater from effluent treatment plant	References
Algal consortium	*Chlorella* sp. *Nitzschia acicularis*	Reactor secondary waste water	Yao et al. (2019)
Bacterial consortium	*Methanosarcina* sp., *Methanotrix* sp., *Methanoculleus* sp., *Methanobacterium* sp., *Methanospirillum* sp.	Complex wastewater from effluent treatment plant	Vassalle et al. (2020)
Algal consortium	*Chlorella vulgarisScenedesmus* sp.. *Westella botryoides*	Sewage waste water	Zhao et al. (2014)
Bacterial consortium	*Acinetobacter junii, Rhodococcus* sp., *Pseudomonas indoloxydans*	Complex wastewater from effluent treatment plant	Boduroglu et al. (2014)
Fungal consortium	Arbuscular Mycorrhizal fungi with *Phragmites australis*	Complex wastewater from effluent treatment plant	Gupta et al. (2014)
Fungal consortium	*Scedosporium apiospermum* and *Aspergillus orchraceus*	Complex wastewater from effluent treatment plant	Santos and Corso (2014)
Bacterial consortium	*Lactobacillus, Saccharomyces, Aspergillus, Pseudomonas, Streptomyces*	Complex wastewater from effluent treatment plant	Adnan et al. (2014)
Bacterial consortium	*Bacillus subtilis, Bacillus Thuringiensis, E. coli, Rhodopseudomonas palustris, Rhodobacter spheroids, Lactobacillus* sp.	Complex wastewater from effluent treatment plant	Hamadi et al. (2017)
Bacterial consortium	*Pseudomonas* sp., *Actinomyceta* sp., *Bacillus* sp.., *Streptomyces* sp.., *Staphylococcus* sp.	Complex wastewater from effluent treatment plant	Marzbali et al. (2017)
Bacterial consortium	*Vibrio, Staphylococcus, Aerococcus, Acinetobacter, Exiguobacterium*	Petrochemical wastewater	Sheekh et al. (2018)
Fungal consortium	*Trametes versicolor, Irpexlacteus, Ganoderma lucidum, Phanerochaetechrysosporium*	Industrial wastewater	Hussein et al. (2018)
Fungal consortium	*Bjerkandera* sp. *R1, Bjerkanderaadusta, Phanerochaetechrysosporium*	Industrial wastewater	Naji and Salman (2019)
Fungi- Algae consortium	*Aspergillus niger* (fungi)*; Chlorella vulgaris* (*algae*)	Pharmaceutical wastewater	Lebron et al. (2019)

(continued)

Table 2.3 (continued)

Consortium species	Genus	Complex wastewater from effluent treatment plant	References
Algae	*Scenedesmus obliquus, Chlorella obliquus*	Pharmaceutical wastewater	Tan et al. (2020)
Algae	*Lessonia nigrescens, Macrocystis integrifolia*	Complex wastewater from effluent treatment plant	Das et al. (2018)
Algae	*Anabaena cylindrica, Spirulina platensis Chlorella, Anabaena, Chlorococcus*	Synthetic wastewater	Das et al. (2018)
Algae	*Haematococcus pluvialis, Chlorella* sp.*, Selenastrumcapricornutum*	Synthetic wastewater	Malik and Ahmed (2012)
Microalgae and Cyanobacteria	*Chlorella* sp.. and *Phormidium* sp.	Tannery wastewater	Meckenstock (2016)

Distributed under the terms of the Creative Commons Attribution License (http://creativecomm ons.org/licenses/by/4.0/) Sharma et al. (2023)

the choice of suitable microbial strains, long-term system modeling and operational parameter optimization, techno-economic viability, etc., should receive more attention.

The current era necessitates the development of economically feasible ecologically friendly technology. The integration of engineered microorganisms from the scientific phase to the practical and pilot phases is essential for achieving substantial advancements in the application of microorganisms during the process of treating wastewater. Future environmental management will require the use of current technologies and efficient collaboration between several disciplines.

References

Adnan LA, MohdYusof AR, Hadibarata T et al (2014) Biodegradation of bis-azo dye reactive black 5 by white-rot fungus *Trametes gibbosa* sp. WRF 3 and its metabolite characterization. Water Air Soil Pollut 225:2119

Ali H, Khan E, Sajad MA (2013) Phytoremediation of heavy metals concepts and applications. Chemosphere 91(7):869–881

Azubuike CC, Chikere CB, Okpokwasili GC (2016) Bioremediation techniques classification based on site of application: principles, advantages, limitations and prospects. World J Microbiol Biotechnol 32(11):180

Barclay M, Hart A, Knowles CJ, Meeussen JC, Tett VA (1998) Biodegradation of metal cyanides by mixed and pure cultures of fungi. Enzym Microb Technol 22(4):223e231

Boduroglu G, Kiliç NK, Onmez GD (2014) Bioremoval of reactive blue 220 by *Gonium* sp. biomass. Environ Technol 35(19):2410–2415

Cao Z, Yan W, Ding M, Yuan Y (2022) Construction of microbial consortia for microbial degradation of complex compounds. Front Bioeng Biotechnol 10:1051233.https://doi.org/10.3389/fbioe.2022.1051233

Das C, Ramaiah N, Pereira E, Naseera K (2018) Efficient bioremediation of tannery wastewater by mono strains and consortium of marine *Chlorella* sp. and *Phormidium* sp.. Int J Phytoremediation 20:284–292

Ganguly D, Prasanna KLV, Neelapu S, Goswami G (2023) Role of microbes in bioremediation. In: Verma P (ed) Industrial microbiology and biotechnology. Springer, Singapore. https://doi.org/10.1007/978-981-99-2816-3_19

Germaine KJ, Byrne J, Liu X, Keohane J, Culhane J, Lally RD, Kiwanuka S, Ryan D, Dowling DN (2015) Ecopiling: a combined phytoremediation and passive biopiling system for remediating hydrocarbon impacted soils at field scale. Front Plant Sci 5:756

Goswami M, Chakraborty P, Mukherjee K, Mitra G, Bhattacharyya P, Dey S, Tribedi P (2018) Bioaugmentation and biostimulation: a potential strategy for environmental remediation. J Microbiol Exp 6:223–231

Gupta VK, Bhushan R, Nayak A et al (2014) Biosorption and reuse potential of a blue green alga for the removal of hazardous reactive dyes from aqueous solutions. Bioremediation J 18(3):179–191

Hamadi AA, Uraz G, Katircioglu H et al (2017). Adsorption of azo dyes from textile wastewater by *Spirulina Platensis*. Eurasian J Environ Res1(1):19–27

Hosseinzadeh A, Baziar M, Alidadi H et al (2020) Application 971 of artificial neural network and multiple linear regression in modeling nutrient recovery in vermicompost 972 under different conditions. Biores Technol 303:122926

Husain R, Vikram N, Yadav G, Kumar D, Pandey S, Patel M, Khan NA, Hussain T (2022) Microorganisms: an eco-friendly tools for the waste management and environmental safety. In: Arora S, Kumar A, Ogita S, Yau YY (eds) Biotechnological innovations for environmental bioremediation. Springer, Singapore. https://doi.org/10.1007/978-981-16-9001-3_36

Hussein MH, Abou El-Wafa GS, Shaaban-Dessuki SA, et al (2018) Bioremediation of methyl orange onto *Nostoccarneum* biomass by adsorption, kinetics and isotherm studies. Global Adv Res J Microbiol 7(1):6–22

Ji J, Kakade A, Yu Z et al (2020) Anaerobic membrane bioreactors for treatment of 991 emerging contaminants: a review. J Environ Manage 2020(270):110913

Kadri T, Rouissi T, Kaur BS et al (2017) Biodegradation of polycyclic aromatic hydrocarbons (PAHs) by fungal enzymes: a review. J Environ Sci 51:52–74

Kang Y, Xu X, Pan H et al (2017) Decolorization of mordant yellow 1 using *Aspergillus* sp. TS-A CGMCC 12964 by biosorption and biodegradation. Bioengineered 9(1):222–232

Kelly DP, Wood AP (2000) Reclassification of some species of *Thiobacillus* to the newly designated genera *Acidithiobacillus* gen. nov., *Halothiobacillus* gen. nov. and *Thermithiobacillus* gen. nov. Int J Syst Evol Microbiol 50(2):511–516

Kong Z, Li L, Kurihara R, Kubota K, Li Y-Y (2018) Anaerobic treatment of N, N-dimethylformamide-containing wastewater by co-culturing two sources of inoculum. Water Res 139:228–239. https://doi.org/10.1016/j.watres.2018.03.078

Kujan P, Prell A, Safár H et al (2006) Use of the industrial yeast *Candida utilis* for cadmium sorption. Folia Microbiol (Praha) 51(4):257–260

Kumar A, Ergas S, Yuan X, Sahu A, Zhang Q, Dewulf J et al (2010) Enhanced CO_2 fixation and biofuel production via microalgae: recent developments and future directions. Trends Biotechnol 28(7):371–380. https://doi.org/10.1016/j.tibtech.2010.04.004

Kumar RK, Agrawal K, Mehariya S et al (2021) Current perspective on wastewater treatment using photobioreactor for *Tetraselmis* sp.: an emerging and foreseeable sustainable approach. Environ Sci Pollut Res 67:1–33

Lebron YAR, Moreira VR, Santos LVS (2019) Studies on dye biosorption enhancement by chemically modified *Fucus vesiculosus, Spirulina maxima* and *Chlorella pyrenoidosa* algae. J Clean Prod 240:118197

Lee Li C, Wang S, Du X et al (2016) Immobilization of iron-and manganese-oxidizing bacteria with a biofilm-forming bacterium for the effective removal of iron and manganese from groundwater. Bioresour Technol 220:76–84

Leong WH, Zaine SNA, Ho YC et al (2019) Impact of various microalgal bacterial populations on municipal wastewater bioremediation and its energy feasibility for lipid-based biofuel production. J Environ Manag 249:109384

Leson G, Winer AM (1991) Biofiltration: an innovative air pollution control technology for VOC emissions. J Air Waste Manag Assoc 41(8):1045–1054

Li M, Wu Y-J, Yu Z-L et al (2007) Nitrogen removal from eutrophic water by floating-bed-grown water spinach (*Ipomoea aquatica* Forsk.) with ion implantation. Water Res 41:3152–3158

Li C, Xiao S, and Ju L-K (2016) Cultivation of phagotrophic algae with waste activated sludge as a fast approach to reclaim waste organics. Water Res 91:195–202. https://doi.org/10.1016/j.wat res.2016.01.021

Li B, Zhang T, and Yang Z (2019) Immobilizing unicellular microalga on pellet-forming filamentous fungus: can this provide new insights into the remediation of arsenic from contaminated water? Bioresour. Technol 284:231–239. https://doi.org/10.1016/j.biortech.2019.03.128

Lim SL, Chu WL, and Phang SM (2010) Use of Chlorella vulgaris for bioremediation of textile wastewater. Biores Technol 101(19):7314–7322.

Marzbali MH, Mir AA, Pazoki M, et al. (2017) Removal of direct yellow 12 from aqueous solution by adsorption onto Spirulina algae as a high-efficiency adsorbent. J Environ Chem Eng 5(2):1946–1956

Meckenstock RU (2016) Anaerobic degradation of benzene and polycyclic aromatic hydrocarbons. J Mol Microbiol Biotechnol 26:92–118

Monica P, Darwin R, Manjunatha B, Zúñiga JJ, Diego R, Bryan RB, Mulla SI, Maddela NR (2016) Evaluation of various pesticides-degrading pure bacterial cultures isolated from pesticide-contaminated soils in Ecuador. Afr J Biotechnol 15:2224–2233

Mrozik A, Piotrowska-Seget Z (2010) Bioaugmentation as a strategy for cleaning up of soils contaminated with aromatic compounds. Microbiol Res 165(5):363–375

Mujtaba G, Rizwan M, Lee K (2017) Removal of nutrients and COD from wastewater using symbiotic co-culture of bacterium Pseudomonas putida and immobilized microalga Chlorella vulgaris. J Industrial Eng Chem 49:145–151. https://doi.org/10.1016/j.jiec.2017.01.021

Nzila A, Razzak SA, Zhu J (2016) Bioaugmentation: an emerging strategy of industrial wastewater treatment for reuse and discharge. Int J Environ Res Publ Health 13(9):846

Oyetibo GO, Miyauchi K, Huang Y, Chien MF, Ilori MO, Amund OO, Endo G (2017) Biotechnological remedies for the estuarine environment polluted with heavy metals and persistent organic pollutants. Int Biodeterior Biodegrad 119:614–625

Parmar S, Daki S, Bhattacharya S, Shrivastav A (2021) Microorganism: an ecofriendly tool for waste management and environmental safety. In: Shah MP, Rodriguez-Couto S, Kapoor RT (eds) Development in wastewater treatment research and processes, chap 8. Elsevier, 2022, pp 175–193. ISBN: 9780323856577. https://doi.org/10.1016/B978-0-323-85657-7.00001-8

Plöhn M, Spain O, Sirin S, Silva M, Escudero-Oñate C, Ferrando-Climent L, et al. (2021) Wastewater treatment by microalgae. Physiol Plant 173(2):568–578. https://doi.org/10.1111/ppl.13427

Raja R, Hemaiswarya S, Kumar NA, Sridhar S, and Rengasamy R (2008) A perspective on the biotechnological potential of microalgae. Crit Rev Microbiol 34(2):77–88. https://doi.org/10.1080/10408410802086783

Ramanan R, Kim B-H, Cho D-H, Oh H-M, and Kim H-S (2016) Algae bacteria interactions: evolution, ecology and emerging applications. Biotechnol Adv 34(1):14–29. https://doi.org/10.1016/j.biotechadv.2015.12.003

Salim S, Kosterink NR, Wacka ND, Vermuë M, and Wijffels RH (2014) Mechanism behind autoflocculation of unicellular green microalgae Ettlia texensis. J Biotechnol, 174:34–8

Sharma M, Agarwal S, Agarwal Malik R, Kumar G, Pal DB, Mandal M, Sarkar A, Bantun F, Haque S, Singh P, Srivastava N, Gupta VK. (2023) Recent advances in microbial engineering approaches for wastewater treatment: a review. Bioengineered. 2023 Dec 14(1):2184518. https://doi.org/10.1080/21655979.2023.2184518

Sheekh MM, Abou-El-Souod GW, AsragHa E. (2018). Biodegradation of some dyes by green algae Chlorella vulgaris and the cyanobacterium Aphano capsaelachista, Egypt J Bot 58(3):311–320

Sierra JDM, Wang W, Cerqueda-Garcia D, Oosterkamp MJ, Spanjers H, and van Lier JB (2018) Temperature susceptibility of a mesophilic anaerobic membrane bioreactor treating saline phenol-containing wastewater. Chemosphere 213:92–102. https://doi.org/10.1016/j.chemosphere.2018.09.023

Subash Chandra Bose SR, Ramakrishnan B, Megharaj M, Venkateswarlu K, and Naidu R (2013) Mixotrophic cyanobacteria and microalgae as distinctive biological agents for organic pollutant degradation. Environ Int 51:59–72. https://doi.org/10.1016/j.envint.2012.10.007

Tan L, Shao Y, Mu G, et al. (2020) Enhanced azo dye biodegradation performance and halotolerance of Candida tropicalis SYF-1 by static magnetic field (SMF). Bioresour Technol 295:122–283

Tyagi M, da Fonseca MM, de Carvalho CC (2011) Bioaugmentation and biostimulation strategies to improve the effectiveness of bioremediation processes. Biodegradation 22(2):231–241

Wani AH, Branion RM, Lau AK (1997) Biofiltration: a promising and cost-effective control technology for Odors, VOCs and air toxics. J Environ Sci Health Part A 32(7):2027–2055

Wijffels RH, Barbosa MJ, Eppink MH (2010) Microalgae for the production of bulk chemicals and biofuels. Biofuel Bioprod Biorefin 4(3):287–295

Xiong J-Q, Miracle MB, and Jeon B-H (2017) Ecotoxicological effects of enrofloxacin and its removal by monoculture of microalgal species and their consortium. Environ Pollut 226:486–493. https://doi.org/10.1016/j.envpol.2017.04.044

Yao S, Lyu S, An Y, et al. (2019) Microalgae–bacteria symbiosis in microalgal growth and biofuel production: a review. J Appl Microbiol 126:359–368.

Zhao X, Chen Z, Wang X, et al. (2014) PPCPs removal by aerobic granular sludge membrane bioreactor. Appl Microbiol Biotechnol 98:9843–9848.

Chapter 3
Wastewater Treatment Using Microorganisms

Abstract Global water usage has nearly doubled, leading to water pollution from factors like urbanization, sewage disposal, industrial processes, population growth, radioactive waste, chemical fertilizers, and pesticides. Wastewater treatment is emerging as a green, sustainable technology to combat water scarcity and ecosystem maintenance. Standard sewage treatment techniques use a consortium of bacterial species, leading to inorganic contaminants and inferior quality sludge. High-rate algal ponds optimize microalgae application, offering economic and ecological benefits. Sustainable wastewater management is crucial for achieving the Sustainable Development Goals.

Keywords Microorganisms · Wastewater treatment · Bacteria · Fungi · Algae · Cyanobacteria · Integrated approach · Sustainable method · High-rate algal ponds · Primary treatment · Secondary treatment · Tertiary treatment · Nanotechnology

3.1 Use of Microorganisms

The volume of water used globally has almost doubled in recent years, which has led to serious worries about water pollution. Numerous factors contribute to this contamination, such as urbanization, sewage disposal, industrial processes, population growth, radioactive waste, chemical fertilizers, and pesticides. Many substances and particles, including microbes, heavy metals, nutrients, and polycyclic aromatic hydrocarbons, may affect the water quality. Organic pollutants have been related to reproductive diseases and cancer, among other major health problems. Wastewater treatment is emerging as a key strategy in the battle against water scarcity and the maintenance of ecosystems. However, traditional wastewater treatment techniques are often expensive, involve intricate infrastructure, and call for a lot of upkeep. As a result, long-term, workable alternatives like microbial wastewater treatment are being investigated as green and sustainable technology. Wastewater can support a

© The Author(s), under exclusive license to Springer Nature Switzerland AG 2025
P. Bajpai, *Developments in Microbial Bioremediation*,
SpringerBriefs in Molecular Science, https://doi.org/10.1007/978-3-031-78319-7_3

Table 3.1 Bioremediation methods and their uses

Techniques	Applications	Effectiveness	Drawbacks	References
Fungal bioremediation	Synthesized effluents	61% COD, 75–89% BOD	Prolonged cycle of spore formation	Bardi et al. (2017), Rana et al. (2017)
Sequence batch reactors	Raw domestic and PIWW mixed wastewater	88% COD, 82% BOD	Not notified	İleri et al. (2003)
Active sludge method	Antibiotics	Not provided total antibiotic concentration		Watkinson et al. (2007)
Bacterial bioremediation	PIWW and condensate cooling water, feed	COD elimination up to 80%	Highly specific	Das et al. (2012), Madukasi et al. (2010)
Up-flow anaerobic stage reactor	Avilamycin, tylosin	70–95%	Not notified	(Chelliapan et al. 2006)
Membrane bioreactors	Thiazide, antibiotics	Not provided	No effect by some drugs	Radjenovic et al. (2007)
Phytoremediation	Tetracycline, phenol, oxytetracycline	57–67%	Not notified	Wang et al. (2015), Gujarathi et al. (2005)
Cyanobacterial bioremediation	Polyethylene with low density	*Nostoc carneum* used 3% carbon from PE	Not notified	Sarmah and Rout (2019)
Phycoremediation	Heavy metals	Above 50% reduction	Not notified	Brar et al. (2017), Yewalkar et al. (2007)

Reproduced with permission from Saeed et al. (2022)

variety of microorganisms that can balance environmental sustainability with socioeconomic requirements, such as bacteria, fungi, and microalgae. An overview of the various bioremediation strategies is provided in Table 3.1.

Current research focuses on isolating various microbes in wastewater treatment systems that have the ability to degrade recently identified contaminants. Further details regarding the significance of key microorganisms in wastewater treatment are covered in the discussion that follows.

3.1.1   Bacteria

Bacteria are the most prevalent microorganism in wastewater treatment systems, and they are essential to these systems. They could be classified as facultatively anaerobic or facultatively aerobic. Both autotrophic and heterotrophic microorganisms are used in wastewater systems however, because heterotrophic bacteria get their energy from carbon-rich organic sources, they are the subject of most research.

Bacteria use the energy they produce by utilizing these organic materials as a substrate to multiply and create new offspring. Water and energy are produced from the organic materials. *Acaligenes* sp, *Arthrobacter* sp., *Achromobacter* sp., *Citromonas* sp., *Pseudomonas* sp., *Flavobacterium* sp., and *Acinetobacter* sp. are a few of the common bacteria identified in wastewater systems. In systems that handle wastewater, bacteria aggregate to create biomass, which promotes biodegradation by assisting in the absorption of more toxins on the surface. Another class of beneficial agents in these wastewater treatment facilities are filamentous bacteria (Mondal and Palit 2019; Parmar et al. 2021; Bhakta et al. 2014). Bacterial cells' capacity to accumulate contaminants, especially metals, is essential for treating wastewater outflow.

The microbial community's makeup and the existence of xenobiotics have an impact on how efficiently biodegradation occurs. Plants supply organic carbon, which encourages rhizosphere microbial activity and speeds up the breakdown of contaminants. Aliphatic aldehydes, amines, and phenols are among the organic molecules that can be broken down by biofilms produced by aquatic plants. Methanotrophic bacteria utilize methane as carbon and energy source while decomposing a range of dangerous organic compounds. Because it produces nitrogen, the aquatic plant *Eichhornia crassipes* has been found to be helpful in the repair of eutrophic waterways. Furthermore, it has been shown that species like *Anabaena oryzae* and *Tolypothrix ceytonica* are excellent at treating industrial effluent. *Plectonema* sp. and *Aphanocapsu* sp. are able to break down crude oil.

Sulfate-reducing bacteria, which include taxa like *Desulfovibrio, Desulfotomaculum, Desulfobacter*, and *Desulfococcus*, are important in the anaerobic sewage treatment process. Bacteria have been recognized as essential biosorbents because to factors including their size, resilience to environmental variations, growth under regulated settings, and density of population. Both active and passive processes can be used by bacterial cells to biosorb metal ions. Passive biosorption can happen in bacterial cells that are not living, whereas living cells take up metal ions by active biosorption. The mechanisms of micro-precipitation, complexation, and chelation facilitate the binding of metal ions. Table 3.2 provides a list of bacteria used in the bioremediation of heavy metals.

3.1.2 Algae

Many kinds of algae, such as *Euglena* sp., *Chlamydomonas* sp., and *Oscillatoria* sp., are frequently used in wastewater treatment (Bhattacharya et al. 2014; Goswami et al. 2021; Darda et al. 2019, Ajayan et al. 2011; Manzoor et al. 2016). Because of its exceptional effectiveness, using algae in treatment procedures has attracted a lot of interest in the recent years. Several bioremediation methods are frequently combined

Table 3.2 Bacteria used to remove heavy metals

Type of heavy metal	Bioremediation using bacterial species	References
Cr	*Acinetobacter* sp.	Magyarosy et al. (2002)
Cr	*Bacillus* sp.	Benazir et al. (2010)
Cr	*Pseudomonas aeruginosa*	Bhargava and Mishra (2018)
Cr	*Cellulosimicrobium* sp. (KX710177)	Salehizadeh and Shojaosadati (2003)
Pb	*Bacillus sfirmus*	Gros et al. (2014)
Pb	*Staphylococcus* sp.	Congeevaram et al. (2007)
Cu, Ni	*Desulfovibriodesulfuricans* KCTC5768 (immobilize on zeolite)	Congeevaram et al. (2007)
Cu, Ni	*Micrococcus* sp.	Jafari et al. (2015)
Co	*Vibriofluvialis*	Al-Garni et al. (2010)
Hg	*Enterobacter cloacae*	Jafari et al. (2015)
Hg	*Klebsiella pneumoniae*	Salehizadeh and Shojaosadati (2003)
Hg	*Bacillus licheniformis*	Badia-Fabregat et al. (2016)
Zn	*Bacillus sfirmus*	Cruz-Morató et al. (2013)
Zn	*Pseudomonas* sp.	Stenholm et al. (2018)
Mn, Zn, Co	*Acetobacter* sp.	Palli et al. (2017)
As	*Herminiimonasarsenicoxydans*	Molla et al. (2002)
Cu, Ni, Cr, U	*Pseudomonas aeruginosa, Aeromonas* sp.	Kang et al. (2018)
Cd	*Bacillus safensis*	Andleeb et al. (2012)
Pb, Cr, Cd	*Aerococcus* sp., *Rhodopseudomonas palustris*	Ibrahim et al. (2018)
Fe	*Microbacterium profundi*	Khan and Fulekar (2017)
Fe, Zn	*Lactobacillus delbrueckii, Streptococcus thermophilus*	Asses et al. (2018)
Mn	*Leptothrix, Pseudomonas, Planctomyces*	Dayi et al. (2020)
Co, Cu, Cr, Pb, Cd	*Lysinibacillus sphaericus, Bacillus safensis*	Sheam et al. (2021)
Pb	*Ralstonia solanacearum*	Gao et al. (2020)
Cd	*Enterobacter aerogenes*	Machado et al. (2010)
Mn, Fe, Cu, U, Zn	*Geobacter* sp., *Pseudomonas fluorescens, Vibrio harveyi, Pseudomonas aeruginosa*	Namara et al. (2008)

Table 3.3 Several species of microalgae utilized in wastewater treatment

Wastewater type	Pollutant	Microalgae	References
Municipal wastewater	Organic waste	*Chlorella vulgaris*	Qi et al. (2021)
Synthetic wastewater	Tetracycline	*Chlorella vulgaris*	Su et al. (2012)
Synthetic wastewater	17 α-Ethynylesteradiol	*Desmodesmussubspicatus*	Du (2021)
Effluent from the wastewater treatment plant	Ciprofloxacin, Progesterone, paracetamol, diclofenac	*Chlorella pyrenoidosa*	Fontalvo et al. (2022)
Pharmaceutical industry wastewater	Salicylic acid, paracetamol	*Chlamydomonas mexicana*	Maqbool et al. (2019)
Municipal wastewater	Organic waste	*Scenedesmus* sp.	Satiro et al. (2022)

with algae to get rid of pesticides, heavy metals, and both organic and inorganic contaminants (Pamar et al. 2021; Mondel and Palit 2019). This covers the application of eukaryotic algae and cyanobacteria in biological wastewater treatment. The use of algal species in bioremediation procedures is referred to as "phycoremediation."

Picochlorum sp., *Chlorella* sp., *Tetraselmis* sp., and *Scenedesmus* sp. are popular cyanobacterial and algal strains used in wastewater treatment. Other strains include *Spirulina* sp., *Chroococcus* sp., *Oscillatoria* sp., *Anabaena* sp., *Scytonema* sp., *Pseudospongiococcus* sp., and *Dolichospermum*. Microalgae are particularly well-suited for wastewater treatment because of their distinctive development characteristics. These traits include their capacity to use the phosphorus, carbon, and nitrogen, present in wastewater in both inorganic and organic forms.

Microalgae have a short life cycle and require little in the way of nutrients. Through the adsorption and desorption processes, algal biomass may be utilized again. Algal biomass can be produced year-round because it can grow regardless of environmental conditions. It can remove heavy metals more effectively than traditional membrane systems. It can act as an oxygen source and help heterotrophic bacteria in the degradation process. In both anaerobic and aerobic treatment systems, algae biomass is advantageous. Microalgae-based wastewater treatment systems can be categorized as either open or closed systems based on the availability of nutrients, amount of money invested, and growing conditions. A selection of different microalgal species utilized in wastewater treatment is detailed in Table 3.3.

3.1.3 Fungi

Numerous fungi can be grown with bacteria and are used in wastewater treatment procedures. There may be rivalry between bacteria and fungus for the available substrates in specific environmental settings. Ammonia can be converted into nitrate by certain species, like *Zooglea* sp., and *Sphaerotilus* sp., which are commonly found in wastewater management systems. Furthermore, during wastewater treatment, to get rid of carbon and nutrition sources, fungi from the genera *Fusarium, Penicillium,* and *Aspergillus,* are frequently employed.

In low-pH settings, when bacterial activity is inhibited, fungi are also essential for breaking down of organic waste. In order to meet their nutritional needs, fungi's hyphae aid in the adsorption and retention of suspended particulate materials. Many fungal species perform a range of enzymatic reactions on different types of substrates during the wastewater treatment process (Mondal and Palit 2019). Since they increase the bioavailability of contaminants and change them into less hazardous forms, fungi are important in the remediation of pollutants, especially heavy metals. They generate a substantial quantity of biomass and are quite simple to grow. Numerous fungal strains have demonstrated to be capable of decomposing a wide range of environmental contaminants, including heavy metals, dyes, and medicinal and aromatic chemicals. Fungi are an excellent option for treating wastewater because they produce a wide variety of extracellular enzymes and have a protective hyphal network that shields their fragile interior organelles from harmful chemicals.

The rhizosphere attracts fungi due to the exudates from the roots. Plants and fungi interact in the rhizosphere due to a variety of circumstances, such as the type of plant, the climate, the quality of the water, the soil, and the existence of other microorganisms. Numerous processes, including denitrification, detoxification, and the release of metal-chelating siderophores, depend on these interactions. Furthermore, fungi have an advantage over bacterial cultures in the treatment of wastewater because they can transform organic waste into biochemicals and other important compounds, especially when it comes to producing proteins and organic acids. Fungal biomass can be used to make animal feed. Numerous fungal species have proven useful in the treatment of wastewater, including *Stachybotrys* sp.*, Pleurotus pulmonarius, Verticillum terrestre, Aspergillus parasitica, Cephalosporium aphidicola, Glomus* sp., *Acremonium* sp.*, Candida* sp.*, Hydnobolites, Minimedusa* sp.*, Peziza, Hydnobolites, Talaromyces,* among others.

Table 3.4 provides an illustration of how fungi effectively break down wastewater. Recent studies have demonstrated the important role lignin degrading fungi play in the breakdown of artificial colors. These fungi generate lignin degrading enzymes that can degrade complex dye structures, like lignin peroxidase, laccase, and manganese peroxidase.

Yeast has been shown in numerous studies to be an efficient way to remove heavy metals and other environmental contaminants. Additionally, because yeast can absorb, collect, and change hazardous chemicals into non-toxic forms, it has been demonstrated to reduce levels of Chemical Oxygen Demand (COD) and remove mono and polyphenols. It has been determined that certain yeast species, including *Trichosporon beigelii, Galactomyces geotrichum, Saccharomyces cerevisiae,* and *Candida krusei,* are efficient at degrading dyes in textile effluent.

Table 3.4 Fungi employed in wastewater pollution remediation

Fungi	Wastewater type	Pollutant	Reference
Penicillium, Saccharomyces	Radioactive waste	Radionuclides U, Th, Sr	Zhou et al. (2014)
Aspergillus niger, Rhizopus oryzae, Saccharomyces cerevisiae, Penicillium chrysogenum	Industrial wastewater	Cr	Hom-Diaz et al. (2015)
Candida sphaerica	Industrial wastewater	Fe, Zn, Pb	Salgueiro et al. (2016)
Candia sp.	Industrial wastewater	Cu, Ni	Fazal et al. (2018)
Candida porapsilosis	Industrial wastewater	Hg	Touliabah et al. (2022)
Sphaerotilus natans	Industrial wastewater	Cr	Das and Chandranand (2011)
Gloeophyllum sepiarium	Industrial wastewater	Cr	Varjani et al. (2020)
Aspergillus niger	Industrial wastewater	Fe	Das and Mishra (2017)
X ray contrast agent iopromide and antibiotic ofloxacin	Hospital wastewater	Trametesversicolor	Asira (2013)
Trametes versicolor	Veterinary hospital wastewater	Ciprofloxacin, trimethoprim, tetracycline, acridone, carbamazepine	Lund et al. (2014)
Trametes versicolor	Urban wastewater	Salicylic acid, Codeine, Ceflalexine, acridone, ciprofloxacine, propanolol	Singh et al. (2022)
Trametes versicolor	Pharmaceutical wastewater	Carbamazepine, benzophenone, diclofenac	Aragaw, (2020)
Pleurotus ostreatus	Hospital wastewater	Diclofenac, ketoprofen, atenolol	Couto et al. (2014)
Penicillium corylophilum	Domestic wastewater	Suspended solids	Abatenah et al. (2017)

Distributed under the terms of the Creative Commons Attribution License from Sharma et al. (2023)

3.1.4   Protozoa

Protozoa are crucial to the methods used to purify wastewater (Madoni 2011). By means of predation, protozoa maintain the density of dispersed bacterial communities, hence increasing the effluent's quality. Although the existence of protozoa was noted practically immediately upon the introduction of each biological wastewater treatment technique, the importance of these microorganisms has only recently come to light. Bacteria are the organisms most directly involved in the treatment of wastewater. They outnumber and outgrow all other groups in terms of biomass, and they have an impact on the mineralization and removal of both organic and inorganic nutrients (Agersborg and Hatfield 1929; Curds and Cockburn 1970a, b; Curds 1975, 1982; Curds and Fey 1969; Curds et al. 1968; Barker 1943). Protozoa like ciliates, flagellates, amoebae, and even small metazoa are frequently found in modern systems with low loads and long sludge retention times. These are eukaryotic organisms that can consume suspended bacteria and other particles. It's commonly believed that their main function in wastewater treatment is to clarify the effluent.

The method of treating wastewater is essentially biological. Microorganisms consume organic materials to remove nutrients from influent as it enters the microbial environment of a treatment plant. Although bacteria remove most nutrients, metazoa and protozoa balance bacterial populations and provide information about wastewater conditions. Operators can cultivate an environment that supports the best possible treatment by being aware of the many functions that wastewater bacteria play and the circumstances that encourage their growth.

Protozoa comprise around 4% of the microbial community in a wastewater system. Protozoa are single-celled microorganisms that are more complicated and larger than bacteria. The protozoa that are most frequently found in wastewater are ciliates, flagellates, and amoeba. Protozoa improve the final effluent's clarity by eating free bacteria and tiny, unsettled floc. Under a microscope, protozoa populations can be used to inform operators about sludge age and treatment conditions.

Amoeba are common before they reach a young sludge age because they need a lot of nutrients or little competition to flourish. Amoeba can also take over in hazardous environments, high particulate matter concentrations, low dissolved oxygen (DO), or shock loads of biochemical oxygen demand (BOD). The last two circumstances typically cause the amoeba to form a gelatinous shell that protects them from other microorganisms. They also use less oxygen, which is essential for growth and reproduction, as they move more slowly. Furthermore, flagellates are typically present till an early age of sludge. Flagellates grow more quickly at the younger sludge age before bacteria have had an opportunity to colonize because they are less competitive with bacteria for the same soluble nutrients. Therefore, a wastewater sample with a relatively high flagellate count may also have a high food to mass (F:M) ratio, which is another term for high-soluble nutrition contents.

Ciliates are important members of aquatic food webs because they eat a variety of protists, bacteria, algae, and even certain metazoans. Ciliates serve as helpful clarifying agents and are good markers of healthy floc production since they feed on bacteria rather than organic waste. In the wastewater microbial ecology, populations of bacteria and algae can proliferate out of control in the absence of ciliates. There are three typical forms of ciliates found in wastewater, and each group has specific conditions that support the growth of its population. As flagellates vanish, swimming ciliates begin to form. When quantities of free bacteria are high enough for predation, their population may rise. A murky effluent may eventually arise from the ciliate population surge if there are excessive amounts of free microorganisms present. When those free bacterial populations start to clump together and form floc through a layer of produced slime, crawling ciliates take over. When the amount of dissolved nutrients is reduced, this slime layer is created. Crawling ciliates start to outcompete swimming ciliates because the latter are less able to detect bacteria present in the floc. Claws that consume germs can enhance flock organization. Stalked ciliates can compete with crawling ciliates at a sludge age that is more developed and has a lower BOD content. Using the cilia surrounding their mouth structure, stalked ciliates anchor themselves to floc and create currents that attract bacteria. Stalked ciliates branch into colonial units in order to more effectively obtain food once their food supplies have dropped noticeably. As sludge ages, the number of stentors and vaginocola protozoa increases.

3.1.5 Metazoa

All animals other than protozoa are classified as metazoa, or multicellular organisms. Metazoa microorganisms play a little role in wastewater treatment; nonetheless, their dominance could suggest the presence of older sludge. Despite their lack of efficacy in purifying wastewater, these larger-than-most protozoa do consume bacteria and feed on algae and protozoa (Calaway 1968). Lagoon treatment systems, which are older systems, are typically where this species predominates. Despite their minor role in the activated sludge system, their existence does provide information on the state of the treatment system. Below are the three most prevalent metazoa that can be found in the activated sludge treatment system:

- Rotifers: They clean up wastewater and are primarily impacted by toxic loads.
- Nematodes: Consume other nematodes as well as bacteria, fungus, and tiny protozoa.
- Tardigrades, or water bears: Able to withstand harsh weather conditions and poisonous sensitivities.
- The primary distinction between metazoa and protozoa is that the former are a class of multicellular animals, while the latter are a group of primitive, unicellular creatures known as protists. Vertebrates (animals having a backbone) and invertebrates (animals lacking a backbone) are the two primary classes of metazoans.

The bilateral body symmetry of metazoans is one of their primary distinguishing characteristics. The ability to move, breathe, consume organic matter, reproduce sexually, and evolve from a hollow ball of cells called a blastula during embryonic development are among the other important characteristics of metazoans.

Nematodes are present in a wide range of wastewater treatment procedures and are frequently seen in systems with higher MCRT levels or in older sludge. When the D.O. concentrations are high and the bacterial food supply is plentiful, they flourish in aerobic processes. Large populations of them can be found in trickling filters, rotating biological contactors (RBCs), secondary wastewater discharge, and places where an older biofilm forms. From an "indicator organism" perspective, they are typically the first to go (in terms of BOD loading) and the last to arrive (in terms of age). They may be indicative of a system with a high MCRT, low food/microorganism ratio, high suspended solids, or an extremely old sludge age. They recycle nutrients and cultivate the bacterial population.

Older activated sludge ages contain bristle worms, which can withstand low-oxygen levels. They can tint a whole clarifier pink and are typically a sign of high nitrate levels. Worms usually point to an older, comparatively healthy system that is not poisonous and is moving toward stability in a sludge-like manner. The system is expected to operate effectively when it is integrated with the correct quantity of stalked ciliates, a limited number of free-swimming ciliates, and a selection of rotifers.

3.2   Combined Use of Bacteria and Microalgae for Wastewater Treatment

Bacteria and algae are widely employed in the process of wastewater treatment, as their growth is affected by factors such as geographical location, the type of pond, the characteristics of the wastewater, and various system parameters. Numerous gram-negative proteobacteria, such as betaproteobacteria, are in charge of eliminating nutrients and organic materials. *Bacteroidetes, Chloroflexi*, and *Acidobacteria*, are among other types. *Trichococcus, Rhodoferax, Rhodobacter, Tetrasphaera, Hyphomicrobium, Candidatus Microthrix* are the most common forms of bacteria that are helpful in the breakdown of microorganisms (Gunther et al. 2020). A cost-effective and environmentally beneficial way to clean wastewater is through microalgae growth, which generates valuable biomass for a variety of uses (Gudiukaite et al. 2021; Alexandre et al. 2018).

Microalgae provides a workable method for treating wastewater since they can remove heavy metals and dangerous organic compounds while utilizing inorganic phosphorus and nitrogen for growth. They serve as an alternative to established methods like activated sludge in the treatment of wastewater. As a byproduct of bacteria in wastewater systems, CO_2 is used by microalgae for photosynthesis, the process by which they make carbohydrates and oxygen. The consortium's makeup influences the absorption of phosphorus, nitrogen, and oxygen as well as CO_2. Reducing the oil and fats that harm the ecosystem and altering the presence of the right bacteria for treating hot or cold water can improve the efficacy of wastewater treatment (Arora et al. 2021; Raouf et al. 2012; Muñoz and Guieysse 2006; Zahara et al. 2020; Dhanker et al. 2021).

By removing organic contaminants and their odor through natural air currents, aerobic sewage treatment plants produce treated effluent that is free of pollutants. Nonetheless, more than 80% of wastewater worldwide is released untreated, wreaking havoc on developing nations. Wastewater treatment plants filter water before returning it to the environment, protecting both ecosystems and people. Significant health concerns are associated with untreated water; each year, it is the cause of 1.7 million deaths, the majority of which take place in poor countries. Waterborne illnesses such as schistosomiasis and cholera are still common, and exposure to heavy metals in the water can cause neurological impairments, cardiovascular problems, kidney damage, cancer, and diabetes.

3.2.1 Strategies for Managing Wastewater Treatment

The four main techniques for treating wastewater are chemical treatment, sludge treatment, biological water treatment, and physical water treatment (Table 3.5).

Traditional Methods of Water Treatment The solids are eliminated using methods like screening, sedimentation, and skimming. This procedure does not involve the use of chemicals. One physical method of treating wastewater is sedimentation, which is mostly utilized to get rid of large or difficult-to-remove particles. When the insoluble material sinks to the bottom, clean water may be separated. Aeration is another practical physical method for purifying water. Air is circulating through the water to provide it oxygen during this process. The final technique, filtration, gets rid of all impurities (Metcalf and Eddy 2003; Mathew and Alan 2009).

Filters that are specifically designed to remove pollutants and insoluble particles from wastewater are used. Sand filters are the most commonly used kind of filters. This method also makes it easy to get rid of oil that builds up on the surface of some effluent (Harleman and Murcott 1999).

Biological Process for Treating Water This breaks down organic materials like oils, soap, food, human waste, that are present in wastewater using a variety of biological processes. Organic compounds are biologically processed by microorganisms found in wastewater.

It is separated into three categories:

(a) Aerobic processes: Organic compounds are broken down by bacteria using oxygen as an oxidizing agent, producing carbon dioxide that plants can ingest.
(b) Anaerobic processes: In these cases, waste is fermented at a particular temperature by fermentation. Anaerobic processes do not require oxygen.
(c) Composting, an aerobic process kind in which sewage is treated by combining it with carbon sources such as sawdust (Arun 2011; Yuansong et al. 2003).

Table 3.5 Techniques for treating wastewater	
	Physical screening, sedimentation, skimming
	Biological breakdown of organic matter
	Chemical chlorine is used as an oxidizing agent
	Sludge involves separation of solid and liquid

3.2.1.1   Chemical Purification of Water

Chemicals are added to the water. An oxidizing substance called chlorine is often used to get rid of the bacteria that contaminate and degrade water (Jari et al. 2003). Ozone serves as an additional oxidizing agent for the treatment of wastewater. To restore the water to its standard pH level of 7, either base or an acid is incorporated into the solution. Water is made pure by adding chemicals that prevent bacteria from growing in it (Ødegaard and Karlsson 1994).

3.2.1.2   Treatment of Sludge

The objectives of this solid–liquid separation method are to eliminate as much leftover water from the solid component and as much solid particle remnants from the liquid component as achievable (Krist et al. 2004; Tong et al. 1980).

3.2.2   Function of Microalgae and Bacteria in Wastewater Treatment

3.2.2.1   Treatment with Microalgae

Aquatic environments primarily produce microalgae, like eukaryotic autotrophic protist-algae and prokaryotic cyanobacteria, which also serve as secondary food sources. Because large-scale algal culture systems are well-understood in terms of biology, ecology, and engineering, algal cultures have been employed for wastewater treatment for over 75 years. By directly absorbing pollutants or converting them into innocuous byproducts, microalgae clean water. They also help bacteria break down organic materials by supplying oxygen through photosynthesis. Microalgae absorb nitrogen, phosphorus, and carbonaceous materials to increase their biomass. Nitrogen is assimilated as ammonia, nitrate, and nitrite.

Given that terrestrial oil seed crops yield less per unit area than microalgae, they can be used to produce biofuel and perhaps remove heavy metals from water. Microalgae use a process known as oxygenic photosynthesis to fix CO_2, which, in the presence of light and water, converts it into reduced carbon molecules like sugars. The wavelength range of photosynthetically active radiation (PAR) is 400–700 nm. For CO_2 fixation or the breaking of two water molecules into one diatomic oxygen molecule, 8 PAR photons of light at 550 nm are needed.

One of the most important components of dry matter of cell is nitrogen, which accounts for 1–10% of it.

Since ammonium prevents other nitrogen sources from being assimilated, it is the preferred nitrogen source for green algae. Nitrate and nitrite are additional nitrogen sources that are crucial for the growth of algae. Their availability is reliant on the functionality and efficiency of cellular systems that transport nitrate and nitrite.

Microalgae use three different processes to remove heavy metals from water:

- Intracellular accumulation;
- Extracellular precipitation;
- Cell surface sorption/complexation.

Phosphorus is used by microalgae to make membrane phospholipids, ATP and nucleic acids. PO_4 and polyphosphate are the types that are most readily absorbed. The use of phosphorus is limited to cyanobacteria. Phosphate-containing organic molecules are hydrolyzed by phosphatases on the membrane surface, whereas Pi transporters carry inorganic phosphate into cells. High light intensity promotes the synthesis of acid soluble polyphosphate and its conversion to proteins and DNA.

Acid insoluble polyphosphate is created from excess phosphorus and is then kept in vacuoles for later use. For the tertiary treatment of wastewater, microalgae such as *C. vulgaris* are frequently employed to extract phosphate and nitrogen from diverse sources.

Because they have a light-dependent, mixotrophic mode of nutrition, they can consume organic carbon as biomass. The microalgae treatment of wastewater is affected by a number of variables:

- Presence of parasites and algae virus organisms;
- High levels of organic substances;
- Cyanobacterial inhibitory substances;
- pH levels;
- O_2 concentrations;
- Temperature;
- Light;
- Availability of CO_2.

The 1950s saw the development of the high-rate algal pond (HRAP) technology, which uses microalgae to clean wastewater and recover resources. Compared to conventional wastewater treatment systems, HRAPs have a greater pathogen treatment and nutrient removal efficiency. The biomass of microalgae can be utilized to produce biofuel, fertilizer, and feed that is high in protein. HRAPs have been utilized for treating wastewater from home, tannery, dairy, and piggery sources. They function best in tropical, semi-arid, or dry environments. Only 20–30% of India's 40 billion liters of wastewater undergo treatment, underscoring the urgent requirement for clean water and sustainable energy solutions in the country. Higher biomass, lipid productivity, and nutrient clearance efficiency are attained by HRAPs. Nevertheless, they are not as competitive as other fuels that are more cost-effective and environmentally friendly. Consequently, HRAPs ought to be taken into account as a competitive substitute for conventional wastewater treatment systems.

3.2.2.2 Treatment with Bacteria

As the main water treatment in the globe, bacteria are essential to the process of treating wastewater. They use organic carbon as a source of nutrients and are kept in tanks with higher concentrations. Activated sludge is a term used to describe the group of microorganisms that remediate wastewater. Three basic processes are used to remove nitrogen from wastewater

- Anaerobic ammonia oxidation (anammox)
- Nitrification
- Denitrification

Various species have various nitrogen metabolisms when using different carbon sources, and nitrogen reduction can be either heterotrophic or autotrophic. Anammox is appropriate for effluents characterized by elevated ammonia levels and reduced carbon content because it eliminates nitrogen without adding any additional organic carbon.

Assimilatory, dissimilatory, and denitrification processes are used to reduce nitrate. Bacteria build up phosphorus through a process known as enhanced biological phosphorus removal (EBPR). While de-polymerization events are a part of degradation processes, bacterial cells use extracellular enzymatic hydrolysis to hydrolyze organic materials in wastewater. Life functions like motility, material transport, ion gradient maintenance, and protein/nucleic acid turnover are powered by this energy. Because biological nutrient removal (BNR) depends on bacterial growth and metabolism, species with higher rates of these two traits are better adapted to this process.

Numerous factors affect the way bacteria treat wastewater. These include the inactivation of anammox bacteria at concentrations of organic matter greater than 300 mg COD/L, the inhibition of anammox activity by methanol, the build-up of glycogen as a result of compositional changes in organic compounds, and the co-transport of phosphorus and potassium. The settling properties of biological solids, gas transfer rates, and microbial activity are all impacted by temperature. The buildup of filamentous organisms lowers the quality of effluent water by preventing sludges from properly compressing and settling. The level of dissolved oxygen fluctuates depending on the treatment solution, while pH plays a significant role in influencing the nitrification process. Through their metabolisms, heterotrophic protozoa can improve nutrient cycling and carbon mineralization by releasing nitrogen as ammonia, nitrate, or phosphates, which speeds up bacteria's uptake of organic carbon. Additionally, heterotrophic protozoa can graze on bacteria, selecting inefficient species for subsequent use. Developed in the 1970s, the up-flow anaerobic sludge blanket (UASB) process treats domestic wastewater and industrial wastewater in tropical to semitropical countries in a cost-effective manner compared to the aerobic process. The UASB reactor achieves volatile fatty acid removal rates between 77 and 79%, along with COD removal efficiencies of 93%. Its optimal performance is observed within the mesophilic to thermophilic temperature range. However, because nutrients are not completely removed, pathogen elimination is only partial and requires post-treatment.

Several parameters significantly affect the ability of microalgae to process wastewater and are closely associated with their growth. Firstly, the availability of CO_2 is crucial; low levels can impede the development of algal biomass, resulting in a reduced rate of nutrient and heavy metal assimilation. Given that atmospheric CO_2 concentrations are substantially below optimal levels at 0.033%, it is advisable to supplement the air with 1–5% CO_2 to promote optimal algal growth (Larsdotter 2006). Secondly, elevated concentrations of ammonium ions in water, particularly at levels around 100 mg/L, can obstruct the photosynthetic processes of microalgae. This leads to the production of ammonia, which can inhibit algal growth when concentrations exceed 30 mg/L and at a pH of 9 (Park and Craggs 2011; Su 2021).

An increase in pH levels results in a reduction of CO_2 absorption, adversely affecting the activity of the RuBisCO enzyme and facilitating the transformation of NH_4^+ into NH_3. Additionally, oxygen concentrations exceeding 20 mg/L trigger photorespiration and the generation of O_2 radicals, which partially inhibit photosynthesis and diminish the growth of microalgae. Furthermore, temperature influences growth by enhancing it up to a certain threshold, beyond which it declines sharply (Larsdotter 2006). While temperature preferences vary by strain, it is generally optimal between 20 and 30 °C. Sustaining cultures within this temperature range enhances their ability to remove nutrients effectively (Salces et al. 2019; de Godos et al. 2017). Illumination promotes the proliferation of microalgae by elevating the rates of photosynthesis to a specific limit, after which it leads to a reduction in photosynthetic efficiency (Park and Craggs 2011). Inhibitory substances produced by cyanobacteria hinder the development of eukaryotic algae (Larsdotter 2006). In addition, the ability of algae to purify wastewater and produce beneficial byproducts is compromised by the existence of parasites and algal viruses, including rotifers and protozoa. Furthermore, acetate poses a risk of internal damage to particular species, as its un-ionized form can cross cell membranes and later ionize inside the cells. Lastly, elevated levels of organic matter obstruct the nutrient absorption capabilities of microalgae (Grobbelaar 1982; Mostert and Grobbelaar 1987; Ogbonna et al. 2000; Larsdotter 2006).

3.2.3 Combined Bacterial-Microalgal Method for Wastewater Treatment

In industrial settings, the bacterial-microalgal strategy is becoming more popular as a means of enhancing the ecological state of water resources, specifically with regard to reducing the levels of nitrogen and phosphorus in wastewater effluents (Ramanan et al. 2016; Jiang et al. 2021; Saravanan et al. 2021; Fallahi et al. 2021; Anand et al. 2023; Zhang et al. 2020; Gonçalves et al. 2019; Quijano et al. 2017;

Fuentes et al. 2016; Su et al. 2012; Safonova et al. 2004). Because microalgae may thrive on organic and inorganic carbon, they can lower the quantities of inorganic nitrogen, phosphorus, and other chemicals found in wastewater. This method for bio-fixing CO_2 is practical and affordable, because it produces biomass from renewable sources.

Through photosynthesis, microalgae can also produce oxygen by utilizing nitrite and nitrates, which are forms of both organic and inorganic nitrogen. The integrated strategy, which requires a single stage of care, simplifies the treatment procedure (Figs. 3.1 and 3.2). A schematic illustration of algal–bacterial systems is shown in Figs. 3.3 and 3.4.

Microalgae release fewer greenhouse gases during treatment than normal algae because they immediately ingest phosphate and ammonia for cell development. The emission factor of the microalgae wastewater treatment process is 0.0047% g N_2O–N g^{-1} N-input. Many wastewater types with varying efficiency have been studied, including agricultural, municipal, refinery, brewery, and industrial effluents.

The strain *Scenedesmus obliquus* has been shown to be effective in removing nutrients (carbon, nitrogen, phosphorous) from piggery effluent (Prandini et al. 2016; Ji et al. 2013). It has been shown that the microalgae species *Chlorella pyrenoidosa* can grow successfully in the effluent of dairy wastewater, lowering chemical properties and mineral concentrations. *C. vulgaris* can remove from brewery effluent 88% of the BOD, 82% of the total dissolved solids (TN), and 54% of the total phosphorus (TP) (Choi 2016). A number of other microalgae species were investigated for their potential in bioremediation, including *Spirulina* sp., *Nannochloropsis* sp., *Chlamydomonas* sp., and *Dunaliella* sp. For the best removal of phosphorus and nitrogen, microalgae must have the right ratio of vital nutrients. When compared to secondary treatment effluent (STE), the removal of phosphorus and inorganic nitrogen from poultry slaughterhouse water (PSW) is more effectively accomplished by *Chlorella* sp., and this is associated with a greater average specific growth rate.

A variety of studies have shown that freshwater microalgae do not adhere to the traditional nitrogen and phosphorus stoichiometry in their growth and capabilities for wastewater treatment. A *C. vulgaris* culture in a microalgal bioreactor removed up to 96.38% of TN and 92.7% of TP from municipal wastewater, demonstrating the effectiveness of using microalgae in PSW treatment through a variety of culturing approaches. Wastewater treatment using algae is a more sustainable method.

Among the benefits of the integrated biological-microalgal method (HRAP) for wastewater treatment include the ability to remove nutrients and organic carbon in a single step, the reduction of the need for mechanical aeration, and the biomass that may be used to synthesize biodiesel or biogas. Nevertheless, it encounters difficulties such the need for light for microalgal growth, the medium's high pH produced by photosynthesis, the growth rate's sluggishness, and the complexity of isolating microalgal biomass.

The design and configuration of the reactor significantly affect the growth of microalgae, which subsequently impacts the wastewater treatment process. Various microalgal approaches can be categorized as open or closed systems, and they can be suspended or immobilized. For wastewater treatment and biomass productivity to

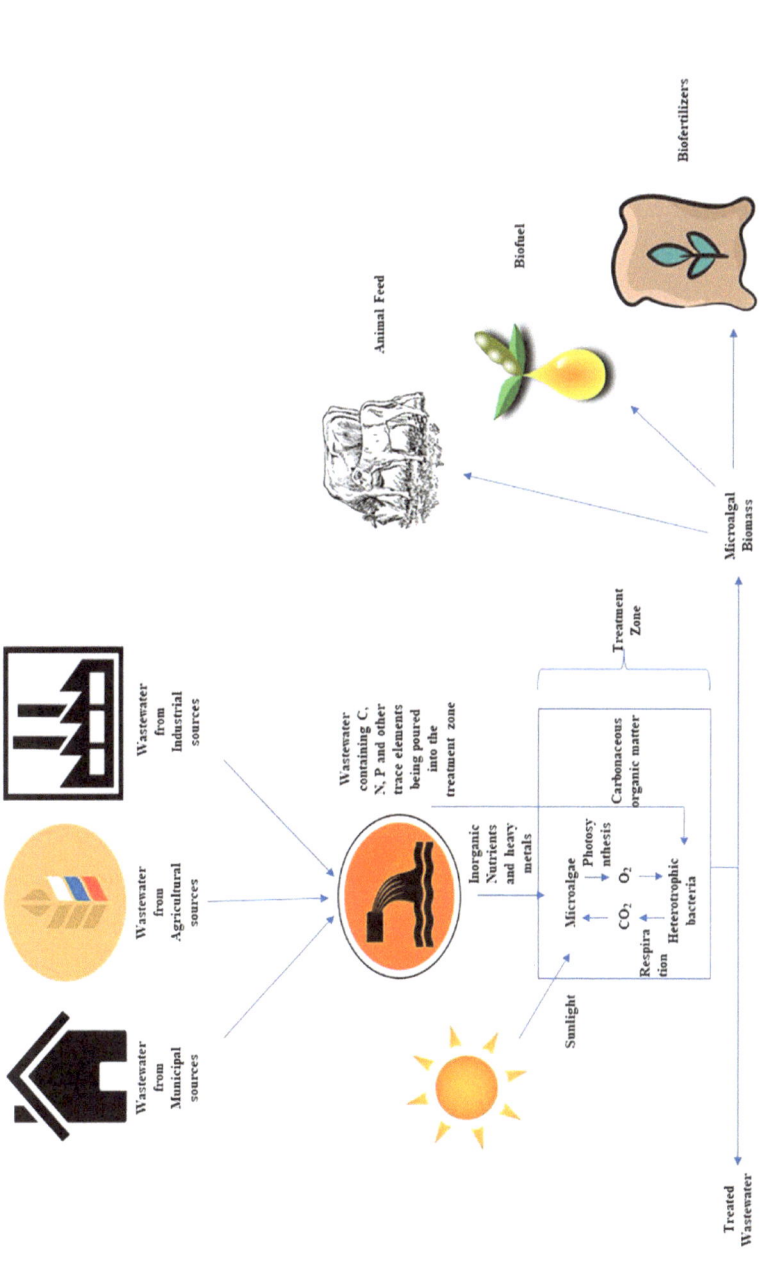

Fig. 3.1 Integrating algae and bacteria for the degradation of wastewater contaminants. Distributed under the terms of the Creative Commons Attribution License (CC BY) from Mathew et al. (2022)

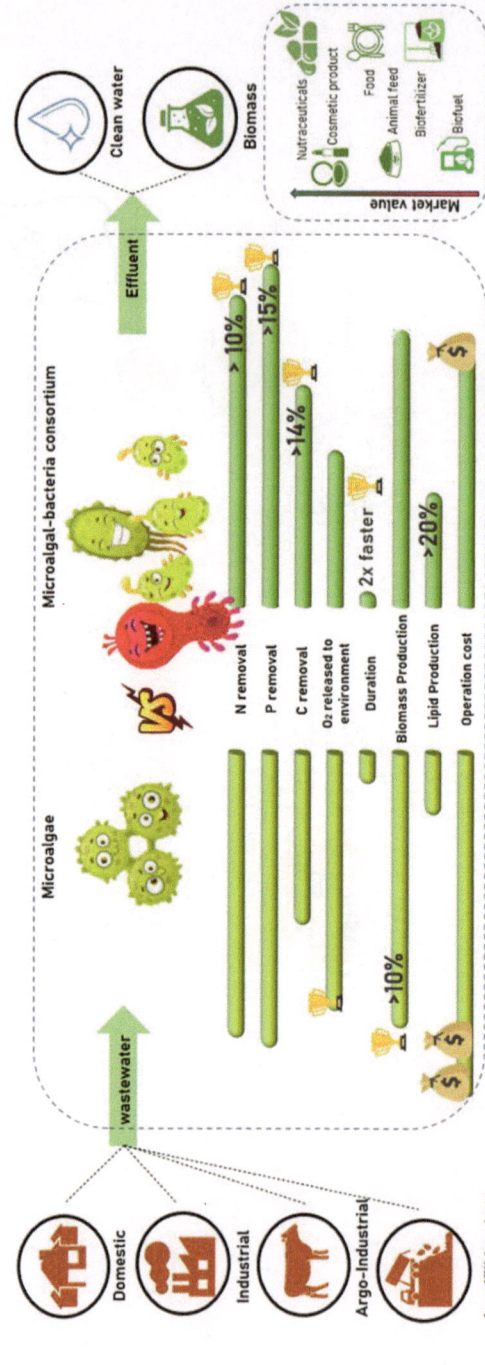

Fig. 3.2 Microalgae-bacteria consortium for wastewater treatment. Reproduced with permission from Aditya et al. (2022)

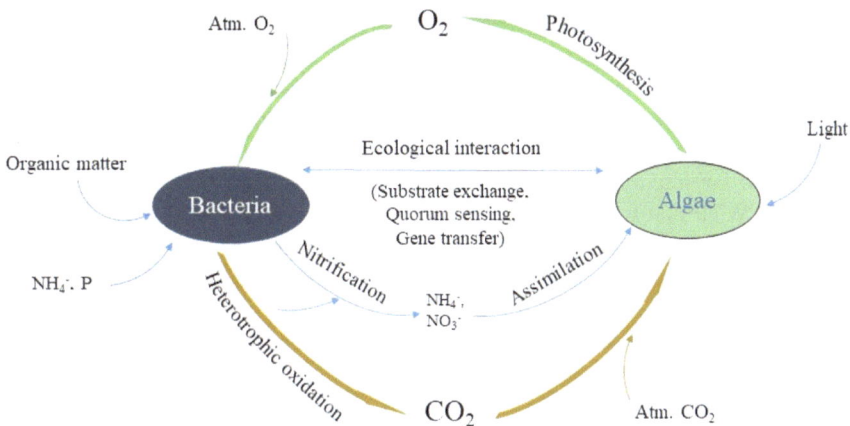

Fig. 3.3 Schematic representation of algal–bacterial symbiosis. Distributed under the terms of the Creative Commons Attribution License from Oruganti et al. (2022)

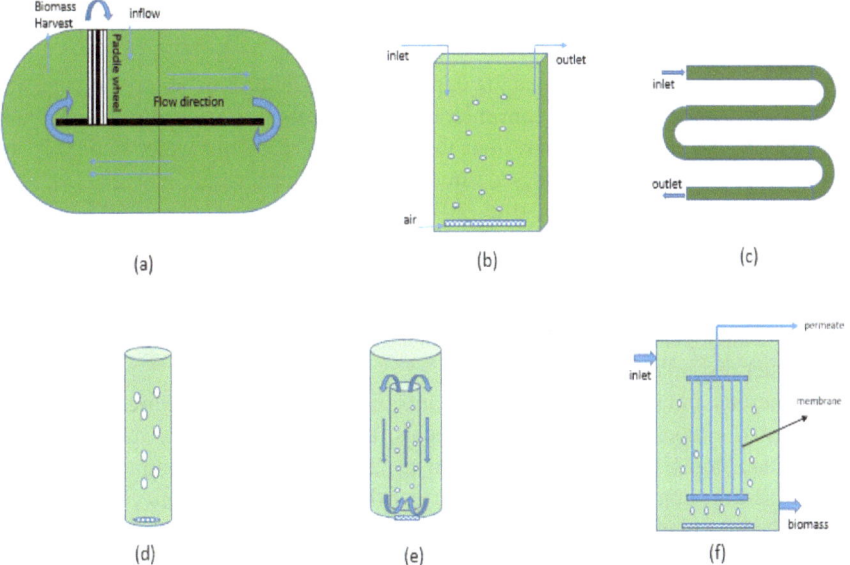

Fig. 3.4 Schematic representation of algal–bacterial systems: **a** high-rate algal pond; **b** flat-plate PBR; **c** tubular PBR; **d** bubble column PBR; **e** internal-looping column PBR; **f** membrane PBR. Distributed under the terms of the Creative Commons Attribution License from Oruganti et al. (2022)

be deemed efficient, the treatment process must achieve the current required effluent concentrations as specified by the urban wastewater treatment directive.

Decentralized treatment of wastewater has become crucial in recent years as centralized wastewater treatment plants struggle to withstand the burden of an ever-increasing metropolitan population and expand their capacity due to space constraints. By integrating algal biofilms into an activated sludge process (ASP)-based DWWTP that was already operating in Hyderabad, India, Katam and Bhattacharyya (2021) studied the integrated strategy. This greatly enhanced the amount of carbon and nutrients that were removed from the influent wastewater in a single reactor.

Table 3.6 shows the ability of microalgae to extract phosphorus and nitrogen from various wastewater sources.

Table 3.7 Illustrates the rate at which certain microalgal species expand under particular conditions by incorporating nutrients into their biomass.

3.3   Approaches for Wastewater Treatment

Water is an essential resource that life depends on, and it needs to be supplied in a way that is sufficient, secure, and convenient. Professionals in wastewater management are responsible for making sure that the quality of receiving water bodies is not jeopardized by wastewater released into the environment (Kesari et al. 2021; Donde Oscar Omondi 2017; Vanessa et al. 2015; Mendes and Domingues 2015). Global environmental sustainability and integrity are seriously threatened by the increasing awareness of the different toxins found in wastewater (Schwarzenbach et al. 2006). The consequences of climate change and population growth have increased the strain on wastewater treatment plants which has led to the depletion and disappearance of freshwater resources (Rop et al. 2014).

Numerous environmental contaminants, including pharmaceuticals, personal care items, endocrine-disrupting chemicals, and other hazardous substances and pathogenic organisms, have been found in many wastewater treatment facilities, raising serious concerns about the quality of the water. Serious concerns are raised by the lack of clarity surrounding the harmful effects of these pollutants (Vaneasa et al. 2015). In order to solve water pollution challenges and ensure the sustainability and safety of aquatic ecosystems, wastewater managers are therefore recommended to implement comprehensive and all-encompassing risk assessment procedures and risk management strategies (Harikishore and Lee 2012; WHO 2011a, b).

There are several instruments available for evaluating sustainability in wastewater treatment. These instruments could include life cycle assessments, economic analyses, and energy flow analyses. Although most assessments concentrate on one tool at a time, a complete evaluation requires the application of a comprehensive collection of indicators due to the complex nature of pollutants in wastewater. But the selection of particular indicators might vary throughout communities due to several variables like population size, culture, and geography. Essential parameters

Table 3.6 Showing effectiveness of different microalgal species at removing total nitrogen (TN) and total phosphorus (TP) from different wastewaters

Main microalgae	Wastewater used	TN initial concentration (mg/L)	TN removed (%)	TP initial concentration (mg/L)	TP removed (%)	References
Micractinium inermum NLP-F014	Bold modified basal freshwater (BMBF) nutrient solution	36 ± 1.1	95.69	49 ± 0.71	10.71	Jeong and Jang (2020)
Scenedesmus dimorphus	Dairy industry wastewater	36.3	> 90	112	20–55	Salces et al. (2019)
Micractinium reisseri	Piggery wastewater	53	7.547	7.1	2.817	Abou-Shanab et al. (2013)
Chlamydomonas sp. (YG04)	Municipal wastewater secondary effluent	190.7 ± 0.12	77.57	19.11 ± 0.03	100	Amini et al. (2014)
Nitzschia cf. *pusila*	Piggery wastewater	53	15.09	7.1	9.859	Abou-Shanab et al. (2013)
Coelastrum microporum	Municipal wastewater	40	88	5.3	89	Salces et al. (2019)
Chlorela zofingiensis	Untreated and Undiluted pig anaerobic digested effluent	1011–1050	82.7	25–26.5	98.17	Li et al. (2019)
Mucidosphaerium pulchellum	Domestic wastewater	64–79	79	4.6–7.2	49	Salces et al. (2019)

Table 3.7 Showing the rate at which different microalgal species grow by assimilating nutrients into their biomass under specific condition sets

Microalgae	Taxonomic order	Biomass productivity as per studies (kg/m^3/day)	References
Chlorala vulgaris	Chlorellales	0.35	Blair et al. (2014)
Chlorala sorakiniala	Chlorellales	0.85	Ugwu et al. (2007)
Chlamyadomonas sp. (YG04)	Chlamydomonadales	0.0552	Shekh et al. (2016)
Coelastrum microporurn	Sphaeropleales	0.044	Hussain et al. (2020)
Oocystis sp.	Chlorellales	0.02524	Mohammadi et al (2018)
Chlorella zofingiensis zofingiensisChroreila zip fingiensis	Chlorellales	0.29616 ± 0.01916	Li et al. (2019)
Scenedes obliquus	Sphaeropleales	0.29250	Ho et al. (2010)
Scenedes dimorphus	Sphaeropleales	0.09	Chng et al. (2017)
Mucidosphaerium pulchellum	Chlorellales	0.1889 ± 0.010	Mehrabadi et al. (2015, 2017)

Distributed under the terms of the Creative Commons Attribution License (CC BY) from Mathew et al. (2022)

for assessing the sustainability of systems for managing and treating wastewater have been discovered by numerous research. While many studies on wastewater sustainability are comprehensive and include a wide range of transdisciplinary indicators, they frequently focus on evaluating a single treatment method without comparing alternatives. This method, which mostly addresses environmental stresses, has a tendency to be one-dimensional. Moreover, research focusing on the economic and environmental components of wastewater treatment often ignores social concerns, resulting in an incomplete understanding of sustainability that takes into account that considers economic, environmental, and social factors.

The selection of treatment technology for management of wastewater requires not only technical considerations but also taking human and environmental interactions into account for comprehensive sustainability. As there is no one-size-fits-all way to treat wastewater, it is not appropriate to focus on just one contaminant. In order to protect public health, proper disinfection must strike a balance between decreasing the production of disinfection byproducts and cutting greenhouse gas emissions.

An unsustainable method of managing wastewater can have effects on future generations as well as the current area of operation. Given this, it is appropriate to alter the standard sustainability measures for wastewater systems, which have mostly ignored social elements in favor of environmental implications, to account for both immediate and long-term repercussions. Moreover, there should be a closer alignment between the design of wastewater management systems and the wider community needs. Local agricultural methods may be more effectively associated with the recycling of treated wastewater and the management of solid residues, thereby redistributing nutrients to the surrounding ecosystem rather than accumulating them in a single water source.

Systems for on-site treatment, including septic tanks, artificial wetlands, and composting latrines, are particularly advantageous for promoting sustainability due to their low energy and chemical requirements, as well as their ability to recycle nutrients back into the ecosystem. A well-balanced assessment of the overall sustainability of certain management strategies with respect to sustainable development will ultimately be necessary to achieve more sustainable wastewater management (Curl et al. 2010; Muga and Mihelcic 2007; Paul and Swaim 2016).

3.3.1 Primary Treatment Process

This preliminary phase of treating wastewater is to remove bulk, suspended, and floating materials. Solids are captured by screening, and suspended debris is eliminated by sedimentation caused by gravity. This process of separating solids from liquids is fundamentally mechanical in nature, although the use of chemicals can sometimes facilitate a quicker sedimentation. The biological oxygen demand (BOD) of the entering wastewater is decreased by 20–30% during this stage of treatment, and the overall amount of suspended particles is decreased by about 50–60%.

3.3.2 Secondary Treatment Process

This phase aids in clearing away any organic trash that has dissolved and gotten beyond the initial barrier. Microbes provide energy, water, and carbon dioxide for their own growth while they consume the organic resources. Additional settling is done after the biological process in order to remove more suspended particles. Secondary treatment is capable of eliminating approximately 85% of suspended solids and BOD. In the course of this operation, carbon-based pollutants accumulate within the secondary settling tank. The sludge possesses the capability to serve as a co-substrate in a biogas plant, thereby facilitating the production of biogas, a blend of CH_4 and CO_2. This biogas can then be harnessed to generate both heat and electricity, thereby enhancing energy distribution. Subsequently, nitrification or denitrification processes are employed to treat the remaining clarified water, effectively removing nitrogen and carbon.

The water within a sedimentation basin is treated further by the introduction of chlorine.

At this juncture, it still harbors a range of chemical, microbiological, and metallic, pollutants. So, for ensuring the water is suitable for repurposing, such as for irrigation, it must first undergo a filtration process before being directed to a disinfection tank. Sodium hypochlorite is employed in this tank for the purpose of disinfecting the wastewater. Upon the completion of this treatment, the water is deemed suitable for irrigation purposes. Moreover, a gravity thickening tank, specifically designed to provide a steady air supply, processes the solid waste generated from the primary and secondary treatment stages. Upon completion of this processing, the solid waste is moved to a dewatering centrifuge tank and thereafter transferred to a lime stabilization tank. Consequently, processed solid waste is generated which can subsequently undergo additional processing to be used as building materials, fertilizer, or landfill.

Apart from the activated sludge process, various other techniques have been developed and are being utilized in large-scale reactors (Kesari et al. 2021). These techniques include trickling filters, ponds (aerobic, facultative, anaerobic, and maturation), artificial wetlands, microbial fuel cells, methanogenic reactors, and anaerobic treatments such as up-flow anaerobic sludge blanket (UASB) reactors. These reactors have long been used in the treatment of wastewater.

Behling et al. (1996) looked at how the UASB reactor functioned in the absence of an outside heat source. Throughout the course of a 200-day experiment, their study maintained a COD loading rate of 1.21 kg $COD/m^3/day$, resulting in an average COD removal rate of 85%.

A combination model utilizing an UASB-Activated Sludge reactor and low-strength wastewater generated from residential sources with a BOD5 level of 340 mg/l was presented by Von-Sperling and Chernicharo in (2005). The findings showed a 40% drop in aeration energy usage and a 60% reduction in sludge output.

Rizvi et al. (2015) used cow dung as an inoculant in the UASB reactor in a different investigation.

3.3.3 Tertiary Treatment Processes

These processes are used when some substances, chemicals, or pollutants remain after the secondary treatment stage. Consequently, processes for tertiary treatment are designed to eliminate roughly 99% of the impurities present in wastewater. The water is treated separately or in conjunction with state-of-the-art techniques like ozone (O_3) exposure, UV radiation, and ultrasonication (US) to make it suitable for human consumption. The elimination of microorganisms and heavy metal pollutants that might still be present in the treated water is the methodology's main objective. To guarantee thorough contamination removal, the water that has undergone secondary treatment is first treated to ultrasonication. This is accomplished by passing through an ozone chamber and being exposed to UV radiation.

The processes through which cells become non-viable during ultrasonication involve free radicals generation and the physical disruption of cellular membranes. The simultaneous use of ultrasound, ultraviolet light, and ozone leads to the generation of free radicals that connect with the membranes of biological contaminants. When there is a disruption in the structural integrity of the cell membrane, chemical oxidants are able to penetrate the cell and inflict damage on internal components. Consequently, ultrasonication, whether applied independently or alongside other techniques, promotes the separation of microbes and improves the efficacy of chemical decontaminants (Phull et al. 1997; Petrier et al. 1992; Hua and Thompson 2000; Kesari et al. 2011a, b; Scherba et al. 1991).

A combined treatment strategy that demonstrated considerable effectiveness in the remediation of textile wastewater was examined by Pesoutova et al. (2011). The utilization of ultrasound technology as an initial treatment, alongside ultraviolet light, has been evaluated (Naddeo et al. 2009; Blume and Neis 2004). Furthermore, a range of combinations involving ultrasound and ultraviolet radiation, in conjunction with TiO2 photocatalysis and ozone, have been evaluated to improve the wastewater disinfection process (Paleologou et al. 2007; Jyoti and Pandit 2004). An essential component of the model proposed by Kesari et al. (2021) for wastewater treatment is the suggestion to evaluate the quality of water after each phase of the treatment process (refer to Fig. 3.5). Once it is confirmed that the appropriate purification standards have been achieved, the processed water may be employed for irrigation, potable purposes, or a range of household uses.

3.3.4 Nanotechnology in Advanced Wastewater Treatment

With advancements in nanotechnology, the application of nanofillers has emerged as a highly efficient approach for the advanced treatment of wastewater. These nanofillers' extremely small hole diameters, ranging from 1 to 5 nm, enable the successful extraction of heavy metals, both inorganic and organic polluting agents, disease-causing microorganisms, and pharmacological active compounds (PhACs) (Vergili 2013; Mohammad et al. 2015). The textile industry has progressively integrated nanofillers for a range of applications, such as bleaching of pulp, pharmaceutical manufacturing, dairy processing, elimination of microbes, and the extraction of heavy metals from wastewater. Water filtration systems was designed by Srivastava et al. (2004) that are both highly efficient and reusable, incorporating carbon nanotubes. These systems have shown exceptional achievement in eliminating pathogenic bacteria such as *Escherichia coli* and *Staphylococcus aureus* and along with Poliovirus sabin-1 from wastewater. Nanofiltration is recognized for its capacity to function at reduced pressures and to utilize less energy than reverse osmosis, all the while achieving superior rejection rates of organic compounds when contrasted with ultrafiltration. Consequently, it represents a suitable option for tertiary wastewater treatment (Abdel-Fatah 2018). In addition to nanofilters, various nanoparticles, including nanoparticles from metal and metal oxide, graphene nanosheets, carbon nanotubes, and polymer-based nanosorbents, can significantly enhance wastewater treatment due to their unique properties.

The performance of several metal oxide nanoparticles was investigated by Kocabas et al. (2012). It was found that nanopowders of ZnO_2, FeO_3, TiO_2, and NiO were particularly effective in eliminating arsenate from wastewater. The problem of cadmium pollution, which carries significant health risks, can be effectively

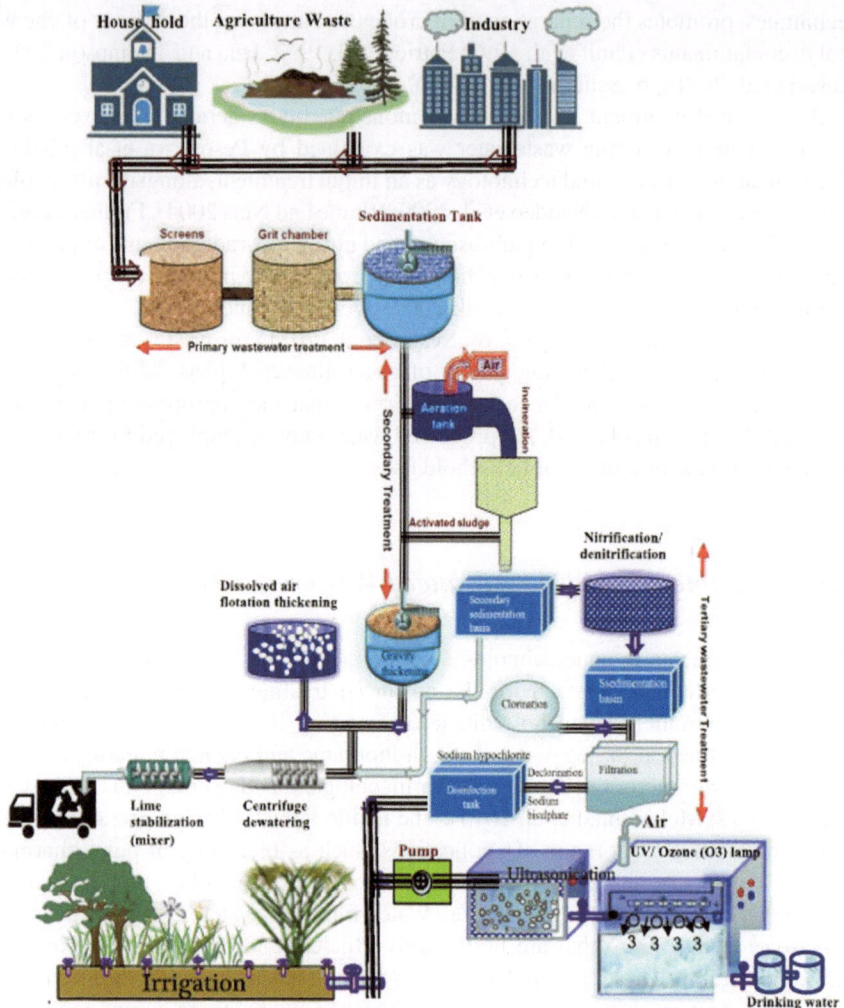

Fig. 3.5 An illustration of the numerous processes used in wastewater treatment that lead to progressively better effluent quality from the source to the treated wastewater's intended usage as irrigation. Distributed under the terms of the Creative Commons CC BY license from Kesari et al. (2021)

addressed through the application of ZnO nanoparticles (Kumar and Chawla 2014). Furthermore, 70% removal rate of mercury was achieved by Vélez et al. (2016).

3.3.5 Wastewater Treatment by Using Plant Species

Various naturally occurring plant species exhibit significant potential as effective agents for the treatment of wastewater, as they possess the ability to absorb pollutants and contaminants, converting them into nutrients (Zimmels et al. 2004; Zhu et al. 2011). Integrating these plant species into management of wastewater systems could provide an economically viable, and operationally simple solution. Additionally, this approach can be implemented in situ, directly at the location where wastewater is generated (Vogelmann et al. 2016).

An evaluation of phytoremediation capabilities of five distinct varieties of plant species: *Eichhornia crassipes, Ipomoea aquatica, Pistia stratiotes, Salvinia molesta*, and *Centella asiatica* was performed by Nizam et al. (2020). They discovered a substantial reduction in phosphate, ammoniacal nitrogen (NH_3–N), and TSS. The removal efficiency for all five species varied impressively from 63.9 to 98% for TSS, NH_3–N, and phosphate.

The physiological impacts of utilizing three prevalent plant species from the Appalachian region for the treatment of household wastewater: common rush (*Juncus effusus* L.), broadleaf cattail (*Typha latifolia* L.), and gray club-rush (*Scirpus validus* L.) were examined by Coleman et al. (2001). Their findings indicated a reduction of approximately 70% in BOD and TSS, alongside a decrease of 50–60% in levels of ammonia, nitrogen, and phosphate, accompanied by a marked decrease in the fecal populations of coliform microorganisms. Zamora et al. (2019) found that the incorporation of three species of ornamental plants—*Hedychium coronarium, Cyperus papyrus*, and *Canna indica*—increased the COD, TSS, N–NO_3, N–NH_4, and P–PO_4 removal effectiveness by 20–60%.

A detailed list of the many plant species employed in the treatment of wastewater may be found in Table 3.8.

3.3.6 Wastewater Treatment by Microorganisms

An array of bacteria, notably *Pseudomonas fluorescens, P. putida*, and *various Bacill*us sp. is integral to the effectiveness of biological systems for treating wastewater. During the course of the treatment procedure, these bacteria work together to build structures like biofilms, flocs, or granules. Moreover, the identification of exopolysaccharides (EPS) produced by bacteria as effective adsorbents introduces an innovative technique for the removal of heavy metals (Gupta and Diwan 2017). Many types of EPS are available commercially. These include alginate, which is obtained from *P. aeruginosa* and *Azotobacter vinelandii*; gellan, which is obtained from *Sphingomonas paucimobilis*; hyaluronan, which is obtained from *Pasteurella multocida, P. aeruginosa*, and some attenuated *Streptococci* strains; xanthan, which is manufactured by *Xanthomonas campestris*; and galactopol, which is obtained from *P. oleovorans* (Freitas et al. 2011a, b, 2009).

Table 3.8 Various plant species applied for the wastewater remediation and their effects

Plant species	Common name	Effects	References
Juncus effusus L.	Common rush or soft rush	Reduction of BOD, COD, TSS	Coleman et al. (2001)
Scirpus validus L.	Grey club-rush		
Typha latifolia L.	Broadleaf cattail or bulrush		
Azolla californiana	Fairy moss	Reduction of turbidity BOD, COD, and TSS	Jacquez and Walner (1985)
Oenanthe javanica	Chinese celery, Indian pennywort, Japanese parsley	Influences dissolved oxygen, pH, and temperature, wastewater purification and nutrient uptake	Zhou and Wang (2010), Zhu et al. (2011)
Hydrocotyle vulgaris	Marsh pennywort	Removal of total nitrogen and NH_4^- nitrogen	Duan et al. (2016)
Ipomoea aquatica	Swamp morning or water spinach		
Eichornia crassipes	Water hyacinth	Reduction of ammonia, nitrate BOD, COD, TSS, turbidity, and heavy metals	Brumer (2000), Jacquez and Walner (1985

Distributed under the terms of the Creative Commons CC BY license from Kesari et al. (2021)

Biological decomposition of urban wastewater using both native and commercial bacterial species, with a particular emphasis on Sludge Hammer was investigated by Hesnawi et al. (2014). The findings revealed a significant reduction in synthetic wastewater, with Sludge Hammer, *B. subtilis*, *B. laterosporus*, and *P. aeruginosa* achieving reductions of 70%, 54%, 52%, and 42%, respectively. The findings reveal that the implementation of bioaugmentation in wastewater treatment reactors, employing both selective and mixed bacterial strains, has the potential to significantly improve treatment efficiency.

In recent times, researchers have shown significant interest in microalgae as a promising alternative treatment system, owing to their efficiency in wastewater management. Algae, which can be either unicellular or multicellular photosynthetic microorganisms, flourish in aquatic environments, saline waters, or moist soils. They utilize excess nutrients such as carbon, phosphorus nitrogen, for their progression and metabolic activities by means of anaerobic processes. This ability of algae also aids in alleviating eutrophication by curbing nutrient overaccumulation in aquatic ecosystems. In the process of nutrient assimilation, algae produce oxygen, thereby enhancing the overall health of the aquatic environment.

References

Abatenah E, Gizaw B, Tsegaye Z et al (2017) The role of microorganisms in bioremediation—a review. J Environ Biol 2(1):038–046

Abdel-Fatah MA (2018) Nanofiltration systems and applications in wastewater treatment: review article. Ain Shams Eng J 9:3077–3092

Abou-Shanab RAI, Ji MK, Kim HC, Paeng KJ, Jeon BH (2013) Microalgal species growing on piggery wastewater as a valuable candidate for nutrient removal and biodiesel production. J Environ Manage 115:257–264. https://doi.org/10.1016/j.jenvman.2012.11.022

Aditya L, Mahlia TMI, Nguyen LN, Vu HP, Nghiem LD (2022) Microalgae-bacteria consortium for wastewater treatment and biomass production. Sci Total Environ 838(Part 1):155871. https://doi.org/10.1016/j.scitotenv.2022.155871

Agersborg HPK, Hatfield WD (1929) The biology of a sewage treatment plant—a preliminary survey. Sew Work J 1:411–415

Ajayan KV, Selvaraju M, Thirugnanamoorthy K (2011) Growth and heavy metals accumulation potential of microalgae grown in sewage wastewater and petrochemical effluents. Pak J Biol Sci 14(16):805–811

Alexandre CL, Antonio V, Diego L, Cristina GF (2018) Energy balance and life cycle assessment of a microalgae-based wastewater treatment plant: a focus on alternative biogas uses. Bioresour Technol 270:138–146. https://doi.org/10.1016/S0960-8524(00)00136

Al-Garni SM, Ghanem KM, Ibrahim AS (2010) Biosorption of mercury by capsulated and slime layer forming Gram-ve *Bacilli* from an aqueous solution. Afr J Biotech 9(38):6413–6421

Amini SR, Najafabady NM, Shaker S, Safari A, Kazemi A, Mousavi P et al (2014) Removal of nitrogen and phosphorus from wastewater using microalgae free cells in bath culture system. Biocatal Agric Biotechnol 3:126–131. https://doi.org/10.1016/j.bcab.2013.09.003

Anand U, Dey S, Parial D et al (2023) Algae and bacteria consortia for wastewater decontamination and transformation into biodiesel, bioethanol, biohydrogen, biofertilizers and animal feed: a review. Environ Chem Lett 21:1585–1609. https://doi.org/10.1007/s10311-023-01562-w

Andleeb S, Atiq N, Robson GD et al (2012) An investigation of anthraquinone dye biodegradation by immobilized *Aspergillus flavus* in fluidized bed bioreactor. Environ Sci Pollut Res 19:1728–1737

Aragaw TA (2020) Functions of various bacteria for specific pollutant degradation and their application in waste water treatment: a review. Int J Environ Sci Technol. https://doi.org/10.1007/s13762-020-03022-2

Arora K, Kaur P, Kumar P, Singh A, Patel SKS, Li X et al (2021) Valorization of wastewater resources into biofuel and value-added products using microalgal system. Front Energy Res 9:646571. https://doi.org/10.3389/fenrg.2021.646571

Arun M (2011) Biological wastewater treatment. Water 33:35–42

Asira E (2013) Factors that determine bioremediation of organic compounds in the soil. Acad J Interdiscip Stud 2:125–128

Asses N, Ayed L, Hkiri N et al (2018) Congo red decolorization and detoxification by *Aspergillus niger*: removal mechanisms and dye degradation pathway. BioMed Res Int 2018. https://doi.org/10.1155/2018/3049686

Badia-Fabregat M, Lucas D, Pereira MA et al (2016) Continuous fungal treatment of non-sterile veterinary hospital effluent: pharmaceuticals removal and microbial community assessment. Appl Microbiol Biotechnol 100:2401–2415

Bardi A, Yuan Q, Tigini V, Spina F, Varese GC, Spennati F et al (2017) Recalcitrant compounds removal in raw leachate and synthetic effluents using the white-rot fungus *Bjerkandera adusta*. Water 9(11):824

Barker AN (1943) The protozoan fauna of sewage disposal plants. Naturalist Hull 806:65–69

Behling E, Diaz A, Colina G, Herrera M, Gutierrez E, Chacin E et al (1996) Domestic wastewater treatment using a UASB reactor. Biores Technol 61:239–245

Benazir JF, Suganthi R, Rajvel D, Pooja MP, Mathithumilan B (2010) Bioremediation of chromium in tannery effluent by microbial consortia. Afr J Biotech 9(21):3140–3143

Bhakta JN, Munekage Y, Ohnishi K et al (2014) Isolation and characterization of cadmium and arsenic-absorbing bacteria for bioremediation. Water Air Soil Pollut 2014(225):2150–2159. https://doi.org/10.1007/s11270-014-2151

Bhargava RN, Mishra S (2018) Hexavalent chromium reduction potential of *Cellulosimicrobium* sp. isolated from common effluent treatment plant of tannery industries. Ecotoxcol Environ Saf 147:102–109

Bhattacharya A, Gupta A, Kaur A, Malik D (2014) Efficacy of *Acinetobacter* sp. B9 for simultaneous removal of phenol and hexavalent chromium from co-contaminated system. Appl Microbiol Biotechnol 98:9829–9841

Blair MH, Kokabian B, Gude VG (2014) Light and growth medium effect on *Chlorella vulgaris* biomass production. J Environ Chem Eng 2:665–674. https://doi.org/10.1016/j.jece.2013.11.005

Blume T, Neis U (2004) Improved wastewater disinfection by ultrasonic pre-treatment. Ultrason Sonochem 11:333–336

Brar A, Kumar M, Vivekanand V, Pareek N (2017) Photoautotrophic microorganisms and bioremediation of industrial effluents: current status and future prospects. 3Biotech 7(1):18

Brumer L (2000) Use of aquatic macrophytes to improve the quality of effluents after chlorination. Ph.D. Dissertation, Technion Israel Institute of Technology, Haifa

Calaway WT (1968) The Metazoa of waste treatment processes-Rotifers. J Water Pollut Control Fed 40(11):R412–R422. http://www.jstor.org/stable/25036170

Chelliapan S, Wilby T, Sallis PJ (2006) Performance of an up-flow anaerobic stage reactor (UASR) in the treatment of pharmaceutical wastewater containing macrolide antibiotics. Water Res 40(3):507–516

Chng LM, Lee KT, Chan DJC (2017) Synergistic effect of pretreatment and fermentation process on carbohydrate-rich *Scenedesmus dimorphous* for bioethanol production. Energy Convers Manag 141:410–419. https://doi.org/10.1016/j.enconman.2016.10.026

Choi H-J (2016) Parametric study of brewery wastewater effluent treatment using *Chlorella vulgaris* microalgae. Environ Eng Res 21:401–408. https://doi.org/10.4491/eer.2016.024

Coleman J, Hench K, Garbutt K, Sexstone A, Bissonnette B, Skousen J (2001) Treatment of domestic wastewater by three plant species in constructed wetlands. Water Air Soil Pollut 3:283–295

Congeevaram S, Dhanarani S, Park J, Dexilin M, Thamaraiselvi K (2007) Biosorption of chromium and nickel by heavy metal resistant fungal and bacterial isolates. J Hazard Mater 146(1–2):270–277. https://doi.org/10.1016/j.jhazmat.2006.12.017

Couto N, Fritt-Rasmussen J, Jensen PE et al (2014) Suitability of oil bioremediation in an Artic soil using surplus heating from an incineration facility. Environ Sci Pollut Res 21:6221–6227

Cruz-Morató C, Ferrando-Climent L, Rodriguez- Mozaz S et al (2013) Degradation of pharmaceuticals in non-sterile urban wastewater by *Trametes versicolor* in a fluidized bed bioreactor. Water Res 47:5200–5210

Curds CR (1975) Protozoa. In: Curds CR, Hawkes HA (eds) Ecological aspects of used-water treatment. Academic Press, London, pp 203–268

Curds CR (1982) The ecology and role of protozoa in aerobic sewage treatment processes. Annu Rev Microbiol 36:27–46

Curds CR, Cockburn A (1970a) Protozoa in biological sewage treatment processes—I. A survey of the protozoan fauna of British percolating filters and activated sludge of sewage and waste treatment plants. Water Res 4:225–236

Curds CR, Cockburn A (1970b) Protozoa in biological sewage treatment processes—II. Protozoa as indicators in the activated sludge process. Water Res 4:237–249

Curds CR, Fey GJ (1969) The effect of ciliated protozoa on fate of *Escherichia coli* in the activated sludge process. Water Res 3:853–867

Curds CR, Cockburn A, Vandyke JM (1968) An experimental study of the role of the ciliated protozoa in the activated sludge process. Water Pollut Control 67:312–329

Curl JM, Ridens MP, Swaim PD, BellamyWD (2010) Water versus energy and greenhouse gases: the numbers that tell the tale. In: Proceedings of the annual conference and exposition. American Water Works Association, Denver, USA

Darda S, Papalas T, Zabaniotou A (2019) Biofuels journey in Europe: currently the way to low carbon economy sustainability is still a challenge. J Clean Prod 208:575–588

Das N, Chandran P (2011) Microbial degradation of petroleum hydrocarbon contaminants: an overview. Biotechnol Res Int 2011:1–13

Das A, Mishra S (2017) Removal of textile dye reactive green-19 using bacterial consortium: process optimization using response surface methodology and kinetics study. J Environ Chem Eng 5(1):612–627

Das MP, Bashwant M, Kumar K, Das J (2012) Control of pharmaceutical effluent parameters through bioremediation. J Chem Pharm Res 4(2):1061–1065

Dayi B, Onac C, Kaya A et al (2020) New type biomembrane: transport and biodegradation of reactive textile dye. J Phys Chem Lett 5(7):9813–9819

Dhanker R, Hussain T, Tyagi P, Singh KJ, Kamble SS (2021) The emerging trend of bio-engineering approaches for microbial nanomaterial synthesis and its applications. Front Microbiol 12:638003. https://doi.org/10.3389/fmicb.2021.638003

Donde OO (2017) Wastewater management techniques: a review of advancement on the appropriate wastewater treatment principles for sustainability. Environ Manag Sustain Dev 6(1):40–58. ISSN 2164-7682

Du H (2021) Water depletion of climax forests over humid karst terrain: patterns, controlling factors and implications. Agric Water Manag. https://doi.org/10.1016/j.agwat.2020.106541

Duan JJ, Zhao JN, Xue LH, Yang LZ (2016) Nutrient removal of a floating plant system receiving low-pollution wastewater: effects of plant species and influent concentration. In: IOP conference series: earth and environmental science, vol 41, no 1

Fallahi A, Rezvani F, Asgharnejad H et al (2021) Interactions of microalgae-bacteria consortia for nutrient removal from wastewater: a review. Chemosphere 272:129878

Fazal T, Mushtaq A, Rehman F, Xu J et al (2018) Bioremediation of textile wastewater and successive biodiesel production using microalgae. Renew Sust Energ Rev 82:3107–3126

Fontalvo NP, Gamero WB, Ardila HA, Gonzalez AF, Ramos CG, Muñoz AE (2022) Removal of nitrogenous compounds from municipal wastewater using a bacterial consortium: an opportunity for more sustainable water treatments. Water Air Soil Pollut 233:1–20

Freitas F, Alves VD, Pais J, Costa N, Oliveira C, Mafra L (2009) Characterization of an extracellular polysaccharide produced by a *Pseudomonas* strain grown on glycerol. Biores Technol 100:859–865

Freitas F, Alves VD, Reis MA (2011a) Advances in bacterial exopolysaccharides: from production to biotechnological applications. Trends Biotechnol 29:388–398

Freitas F, Alves VD, Torres CA, Cruz M, Sousa I, Melo MJ (2011b) Fucose-containing exopolysaccharide produced by the newly isolated Enterobacter strain A47 DSM 23139. Carbohyd Polym 83:159–165

Fuentes JL, Garbayo I, Cuaresma M, Montero Z, González-Del-Valle M, Vílchez C (2016) Impact of microalgae-bacteria interactions on the production of algal biomass and associated compounds. Mar Drugs 14:100

Gao T, Qin D, Zuo S et al (2020) Decolorisation and detoxification of triphenylmethane dyes by isolated endophytic fungus, *Bjerkanderta adusta* SWUSI4 under non nutritive conditions. Bioresour Bioprocess 7(53). https://doi.org/10.1186/s40643-020-00340-8

de Godos I, Arbib Z, Lara E, Cano R, Munoz R, Rogalla F (2017) Wastewater treatment in algal systems, in innovative wastewater treatment and resource recovery technologies. In: Lema JM, Suarez S (eds) Impacts on energy, economy and environment. IWA Publishing, London

Gonçalves AL, Santos FM, Pires JCM (2019) Microalgal consortia: from wastewater treatment to bioenergy production. In: Hallmann A, Rampelotto PH (eds) Grand challenges in algae biotechnology. Springer International Publishing, Cham, Switzerland, pp 371–398

Goswami RK, Agrawal K, Verma P (2021) Microalgae-based biofuel-integrated biorefinery approach as sustainable feedstock for resolving energy crisis. In: Srivastava V (ed) Bioenergy research: commercial opportunities and challenges. Clean Energy Production Technologies. Springer, Singapore. pp 267–293

Grobbelaar JU (1982) Potential of algal production. Water SA 8:79–85. https://doi.org/10.1016/0144-4565(90)90079-Y

Gros M, Cruz-Morato C, Marco-Urrea E et al (2014) Biodegradation of the X-ray contrast agent iopromide and the fluoroquinolone antibiotic ofloxacin by the white rot fungus *Trametes versicolor* in hospital wastewaters and identification of degradation products. Water Res 60:228–241

Gudiukaite R, Nadda AK, Gricajeva A, Shanmugam S, Nguyen DD, Lam SS (2021) Bioprocesses for the recovery of bioenergy and value added products from wastewater: a review. J Environ Manage 300:113831. https://doi.org/10.1016/j.jenvman.2021.113831

Gujarathi NP, Haney BJ, Park HJ, Wickramasinghe SR, Linden JC (2005) Hairy roots of *Helianthus annuus*: a model systemto study phytoremediation of tetracycline and oxytetracycline. Biotechnol Prog 21(3):775–780

Gunther S, Trutnau M, Kleinsteuber S, Hause G, Bley T, Röske I et al (2020) Dynamic of polyphosphate-accumulating bacteria in wastewater treatment plant microbial communities detected via DAPI (40,60-Diamidino-2-Phenylindole) and tetracycline labeling. Appl Environ Microbiol 75:2111–2121. https://doi.org/10.1128/AAC.00345-16

Gupta P, Diwan B (2017) Bacterial exopolysaccharide mediated heavy metal removal: a review on biosynthesis, mechanism and remediation strategies. Biotechnol Rep 13:58–71

Harikishore KRD, Lee SM (2012) Water pollution and treatment technologies. Environ Anal Toxicol 2(5):1000–1003

Harleman DRF, Murcott S (1999) The role of physical-chemical wastewater treatment in the megacities of the developing world. Water Sci Technol 40:75–80. https://doi.org/10.5897/AJB

Hesnawi R, Dahmani K, Al-Swayah A, Mohamed S, Mohammed SA (2014) Biodegradation of municipal wastewater with local and commercial bacteria. Procedia Eng 70:810–814

Ho SH, Chen WM, Chang JS (2010) *Scenedesmus obliquus* CNW-N as a potential candidate for CO_2 mitigation and biodiesel production. Bioresour Technol 101:8725–8730. https://doi.org/10.1016/j.biortech.2010.06.112

Hom-Diaz A, Llorca M, Rodríguez-Mozaz S, et al. (2015). Microalgae cultivation on wastewater digestate: β-estradiol and 17α-ethynylestradiol degradation and transformation products identification. J Environ Manag155:106–113

Hua I, Thompson JE (2000) Inactivation of *E. coli* by sonication at discrete ultrasonic frequencies. Water Res 34:3888–3893

Hussain J, Wang X, Sousa L, Ali R, Rittmann BE, Liao W (2020) Using non-metric multi-dimensional scaling analysis and multi-objective optimization to evaluate green algae for production of proteins, carbohydrates, lipids, and simultaneously fix carbon dioxide. Biomass Bioenergy 141:105711. https://doi.org/10.1016/j.biombioe.2020.105711

Ibrahim NN, Talib SA, Ismail HN et al (2018) Decolorization of reactive red-120 by using macrofungus and microfungus research article special issue. J Fundam Appl Sci 9(6S):954–964

İleri R, Sengil IA, Kulac S, Damar Y (2003) Treatment of mixed pharmaceutical industry and domestic wastewater by sequencing batch reactor. J Environ Sci Health A 38(10):2101–2111

Jacquez RB, Walner HZ (1985) Combining nutrient removal with protein synthesis using a water hyacinth-freshwater prawn polyculture wastewater treatment system. New Mexico Water Resources Research Institute, p 92

Jafari SA, Cheraghi S, Mirbakhsh M, Mirza R, Maryamabadi A (2015) Employing response surface methodology for optimization of mercury bioremediation by *Vibrio parahaemolyticus* PG02 in coastal sediments of Bushehr, Iran. CLEAN–Soil Air Water 43(1):118–126. https://doi.org/10.1002/clen.201300616

Jari K, Anja S, Helvi HT (2003) Elimination of enteric bacteria in biological—chemical wastewater treatment and tertiary filtration units. Water Res 37:690–698. https://doi.org/10.1016/s0043-1354(02)00305-6

Jeong D, Jang A (2020) Exploration of microalgal species for simultaneous water treatment and biofuel production. Environ Res 188:109772. https://doi.org/10.1016/j.envres.2020.109772

Ji M, Abou-Shanab R, Hwang J, Timmes T, Kim H, Oh Y et al (2013) Removal of nitrogen and phosphorus from piggery wastewater effluent using the green microalga *Scenedesmus obliquus*. Environ Eng 139:1198–1205. https://doi.org/10.1061/(ASCE)EE.1943-7870.0000726

Jiang L, Li Y, Pei H (2021) Algal–bacterial consortia for bioproduct generation and wastewater treatment. Renew Sustain Energy Rev 149:111395

Jyoti KK, Pandit AB (2004) Ozone and cavitation for water disinfection. Biochem Eng J 38:2249–2258

Kang Y, Xu X, Pan H et al (2018) Decolorization of mordant yellow 1 using *Aspergillus* sp. TS-A CGMMC 12964 by biosorption and biodegradation. Bioengineered 9(1):222–232

Katam K, Bhattacharyya D (2021) Improving the performance of activated sludge process with integrated algal biofilm for domestic wastewater treatment: system behavior during the start-up phase. Bioresour Technol Rep 13:100618. https://doi.org/10.1016/j.biteb.2020.100618

Kesari KK, Verma HN, Behari J (2011b) Physical methods in wastewater treatment. Int J Environ Technol Manage 14:43–66

Kesari KK, Soni R, Jamal QMS et al (2021) Wastewater treatment and reuse: a review of its applications and health implications. Water Air Soil Pollut 232:208. https://doi.org/10.1007/s11270-021-05154-8

Kesari KK, Kumar S, Verma HN, Behari J (2011a) Influence of ultrasonic treatment in sewage sludge. Hydrol: Curr Res 2:115

Khan R, Fulekar MH (2017) Mineralization of a sulfonated textile dye Reactive Red 31 from simulated wastewater using pellets of *Aspergillus bombycis*. Bioresour Bioprocess 4(1):23

Kocabas ZO, Aciksoz B, Yurum Y (2012) Binding mechanisms of As(III) on activated carbon/titanium dioxide nanocomposites: a potential method for arsenic removal from water. In: MRS online proceedings library, vol 1449. Cambridge University Press

Krist VG, Mark VL, Mogens H, Morten L, Sten BJ (2004) Activated sludge wastewater treatment plant modelling and simulation: state of the art. Environ Model Softw 19:763–783. https://doi.org/10.1016/j.envsoft.2003.03.005

Kumar R, Chawla J (2014) Removal of cadmium ion from water/wastewater by nano-metal oxides. Water Qual Expo Health 5:4

Larsdotter K (2006) Wastewater treatment with microalgae—a literature review. Vatten 62:31–38

Li G, Bai X, Li H, Lu Z, Zhou Y, Wang Y et al (2019) Nutrients removal and biomass from anaerobic digested effluent by microalgae: a review. Int J Agric Biol 12:8–13. https://doi.org/10.25165/j.ijabe.20191205.3630

Lund P, Tramonti A, De Biase D (2014) Coping with low pH: molecular strategies in neutralophilic bacteria. FEMS Microbiol Rev 38:1091–1125

Machado MD, Soares EV, Soares HM (2010) Removal of heavy metals using a brewer's yeast strain of *Saccharomyces cerevisiae*: chemical speciation as a tool in the prediction and improving of treatment efficiency of real electroplating effluents. J Hazard Mater 180(1–3):347–353

Madoni P (2011) Protozoa in wastewater treatment processes: a minireview. Ital J Zool 78(1):3–11. https://doi.org/10.1080/11250000903373797

Madukasi EI, Dai X, He C, Zhou J (2010) Potentials of phototrophic bacteria in treating pharmaceutical wastewater. Int J Environ Sci Technol 7(1):165–174

Magyarosy A, Laidlaw R, Kilaas R et al (2002) Nickel accumulation and nickel oxalate precipitation by *Aspergillus niger*. Appl Microbiol Biotechnol 59(2–3):382–388

Manzoor M, Ma R, Shakir HA et al (2016) Microalgal-bacterial consortium: a cost-effective approach of wastewater treatment in Pakistan, Punjab Univ. J Zool 2016(31):307–320

Maqbool M, Bhatti HN, Sadaf S, Zahid M, Shahid M (2019). A robust approach towards green synthesis of polyaniline-Scenedesmus biocomposite for waste water treatment applications. Mater Res Express 6(5)

Mathew MZ, Alan HM (2009) High quality reuse water by chemical physical wastewater treatment. J Water Contr Pollut 42:437–463. https://doi.org/10.28991/esj-2019-01187

Mathew MM, Khatana K, Vats V, Dhanker R, Kumar R, Dahms H-U, Hwang J-S (2022) Biological approaches integrating algae and bacteria for the degradation of wastewater contaminants—a review. Front Microbiol 12:801051. https://doi.org/10.3389/fmicb.2021.801051

Mehrabadi A, Craggs R, Farid MM (2015) Wastewater treatment high rate algal ponds (WWTHRAP) for low-cost biofuel production. Bioresour Technol 184:202–214. https://doi.org/10.1016/j.biortech.2014.11.004

Mehrabadi A, Farid MM, Craggs R (2017) Potential of five different isolated colonial algal species for wastewater treatment and biomass energy production. Algal Res 21:1–8. https://doi.org/10.1016/j.algal.2016.11.002

Mendes DS, Domingues S (2015) On the track for an efficient detection of *Escherichia coli* in water: a review on PCR-based methods. Ecotoxicol Environ Saf 113:400–411

Metcalf E, Eddy H (2003) Wastewater engineering: treatment and reuse, 4th edn. McGraw-Hill Higher Education, Columbus

Mohammad AW, Teow YH, Ang WL, Chung YT, Oatley-Radcliffe DL, Hila N (2015) Nanofiltration membranes review: recent advances and future prospects. Desalination 356:226–254

Mohammadi M, Mowla D, Esmaeilzadeh F, Ghasemi Y (2018) Cultivation of microalgae in a power plant wastewater for sulfate removal and biomass production: a batch study. J Environ Chem Eng 6:2812–2820. https://doi.org/10.1016/j.jece.2018.04.037

Molla AH, Fakhru'l-Razi A, Hanafi MM et al (2002) Potential nonphytopathogenic filamentous fungi for bioconversion of domestic wastewater sludge. J Environ Sci Heal Part A 37(8):1495–1507

Mondal S, Palit D (2019) Effective role of microorganism in waste management and environmental sustainability. In: Jhariya M, Banerjee A, Meena R, Yadav D (eds) Sustainable agriculture, forest and environmental management. Springer, Singapore. https://doi.org/10.1007/978-981-13-6830-1_14

Mostert ES, Grobbelaar JU (1987) The influence of nitrogen and phosphorus on algal growth and quality in outdoor mass algal cultures. Biomass 13:219–233. https://doi.org/10.1016/0144-4565(87)90061-8

Muga EH, Mihelcik RJ (2007) Sustainability of wastewater treatment technologies. J Environ Manage 88(3):437–447

Muñoz R, Guieysse B (2006) Algal-bacterial processes for the treatment of hazardous contaminants: a review. Water Res 40:2799–2815. https://doi.org/10.1016/j.watres.2006.06.011

Naddeo V, Belgiorno V, Ricco D, Kassinos D (2009) Degradation of diclofenac by sonolysis, ozonation and their simultaneous application. Ultrason Sonochem 16:790–794

Namara CJ, Anastasiou CC, O'flaherty V et al (2008) Bioremediation of olive mill wastewater. Int Biodeterior Biodegrad 61(2):127–134

Nizam NUM, Hanafiah MM, Noor IM, Karim HIA (2020) Efficiency of five selected aquatic plants in phytoremediation of aquaculture wastewater. Appl Sci 10:2712

Ødegaard H, Karlsson I (1994) Chemical wastewater treatment—value for money. In: Klute R, Hahn HH (eds) Chemical water and wastewater treatment III. Springer, Berlin. https://doi.org/10.1007/978-3-642-79110-9_14

Ogbonna JC, Yoshizawa H, Tanaka H (2000) Treatment of high strength organic wastewater by a mixed culture of photosynthetic microorganisms. J Appl Phycol 12:277–284. https://doi.org/10.1023/A:1008188311681

Oruganti RK, Katam K, Show PL, Gadhamshetty V, Upadhyayula VKK, Bhattacharyya D (2022) A comprehensive review on the use of algal-bacterial systems for wastewater treatment with emphasis on nutrient and micropollutant removal. Bioengineered 13(4):10412–10453. https://doi.org/10.1080/21655979.2022.2056823

Paleologou A, Marakas H, Xekoukoulotakis NP, Moya A, Vergara Y, Kalogerakis N, Gikas P, Mantzavinos D (2007) Disinfection of water and wastewater by TiO$_2$ photocatalysis, sonolysis and UV-C irradiation. Catal Today 129:136–142

Palli L, Castellet-Rovira F, Péerez-Trujillo M et al (2017) Preliminary evaluation of *Pleurotus ostreatus* for the removal of selected pharmaceuticals from hospital wastewater. Biotechnol Prog 33:1529–1537

Park JBK, Craggs RJ (2011) Nutrient removal in wastewater treatment high rate algal ponds with carbon dioxide addition. Water Sci Technol 63:1758–1764. https://doi.org/10.2166/wst.2011.114

Parmar S, Daki S, Bhattacharya S, Shrivastav A (2021) Microorganism: an ecofriendly tool for waste management and environmental safety. In: Shah MP, Rodriguez-Couto S, Kapoor RT (eds) Development in wastewater treatment research and processes, chap 8. Elsevier 2022, pp 175–193. ISBN 9780323856577. https://doi.org/10.1016/B978-0-323-85657-7.00001-8

Paul D, Swaim PE (2016) Sustainability principles for drinking water and reuse treatment. In: Symposium for sustainable infrastructure, Colorado

Pesoutova R, Hlavinek P, Matysikova J (2011) Use of advanced oxidation processes for textile wastewater treatment—a review. Food Environ Saf 10:59–65

Petrier C, Jeunet A, Luche JL, Reverdy G (1992) Unexpected frequency effects on the rate of oxidative processes induced by ultrasound. J Am Chem Soc 25:148–3150

Phull SS, Newman AP, Lorimer JP, Pollet TJ, Mason TJ (1997) The development and evaluation of ultrasound in the biocidal treatment of water. Ultrason Sonochem 4:157–164

Prandini JM, da Silva MLB, Mezzari MP, Pirolli M, Michelon W, Soares HM (2016) Enhancement of nutrient removal from swine wastewater digestate coupled to biogas purification by microalgae *Scenedesmus* spp. Bioresour Technol 202:67–75. https://doi.org/10.1016/j.biortech.2015.11.082

Qi F, Jia Y, Mu R et al (2021) Convergent community structure of algal bacterial consortia and its effect on advanced wastewater treatment and biomass production. Sci Rep 11:21118

Quijano G, Arcila JS, Buitrón G (2017) Microalgal-bacterial aggregates: applications and perspectives for wastewater treatment. Biotechnol Adv 35:772–781

Radjenovic J, Petrovic M, Barceló D (2007) Analysis of pharmaceuticals in wastewater and removal using a membrane bioreactor. Anal Bioanal Chem 387(4):1365–1377

Ramanan R, Kim BH, Cho DH et al (2016) Algae-bacteria interactions: evolution, ecology and emerging applications. Biotechnol Adv 34:14–29

Rana RS, Singh P, Kandari V, Singh R, Dobhal R, Gupta S (2017) A review on characterization and bioremediation of pharmaceutical industries' wastewater: an Indian perspective. Appl Water Sci 7(1):1–2

Raouf NA, Homaidan AAA, Ibraheem IBM (2012) Microalgae and wastewater treatment. Saudi J Biol Sci 19:257–275. https://doi.org/10.1016/j.sjbs.2012.04.005

Rizvi H, Ahmad N, Abbas F, Bukhari IH, Yasar A, Ali S, Yasmeen T, Riaz M (2015) Start-up of UASB reactors treating municipal wastewater and effect of temperature/sludge age and hydraulic retention time (HRT) on its performance. Arab J Chem 8:780–786

Rop KR, Muia AW, Makindi S, Donde OO (2014) Changes in the densities of faecal and organic matter contaminants from upstream to downstream along Nyangores river of Mara Catchment, Kenya. J Environ Sci Water Res 3(1):015–25

Saeed MU, Hussain N, Sumrin A, Shahbaz A, Noor S, Bilal M, Aleya L, Iqbal HMN (2022) Microbial bioremediation strategies with wastewater treatment potentialities—a review. Sci Total Environ 818:151754. https://doi.org/10.1016/j.scitotenv.2021.151754

Safonova E, Kvitko KV, Iankevitch MI, Surgko LF, Afti IA, Reisser W (2004) Biotreatment of industrial wastewater by selected algal-bacterial consortia. Eng Life Sci 4:347–353

Salces BM, Riano B, Hernandez D, Gonzalez MCG (2019) Microalgae and wastewater treatment: advantages and disadvantages. In: Alam M, Wang Z (eds) Microalgae biotechnology for development of biofuel and wastewater treatment. Springer, Singapore. https://doi.org/10.1007/978-981-13-2264-8_20

Salehizadeh H, Shojaosadati SA (2003) Removal of metal ions from aqueous solution by polysaccharide produced from *Bacillus firmus*. Water Res 37(17):4231–4235. https://doi.org/10.1016/S0043-1354(03)00418-4

Salgueiro JL, Pérez L, Maceiras R et al (2016) Bioremediation of wastewater using *Chlorella vulgaris* microalgae: phosphorus and organic matter. Int J Environment Res 10:465–470

Saravanan A, Kumar PS, Varjani S et al (2021) A review on algal-bacterial symbiotic system for effective treatment of wastewater. Chemosphere 271:129540

Sarmah P, Rout J (2019) Cyanobacterial degradation of low-density polyethylene (LDPE) by *Nostoc carneum* isolated from submerged polyethylene surface in domestic sewage water. Energ Ecol Environ 4:240–252. https://doi.org/10.1007/s40974-019-00133-6

Satiro J, Cunha A, Gomes AP et al (2022) Optimization of microalgae-bacteria consortium in the treatment of paper pulp waste water. Appl Sci 12:5799

Scherba G, Weigel RM, Obrien WD (1991) Quantitative assessment of the germicidal efficacy of ultrasonic energy. Appl Environ Microbiol 57:2079–2084

Schwarzenbach RP, Escher BI, Fenner K, Hofstetter TB, Johnson CA et al (2006) The challenge of micropollutants in aquatic systems. Science 313:1072–1077

Sharma M, Agarwal S, Agarwal Malik R, Kumar G, Pal DB, Mandal M, Sarkar A, Bantun F, Haque S, Singh P, Srivastava N, Gupta VK (2023) Recent advances in microbial engineering approaches for wastewater treatment: a review. Bioengineered 14(1)

Sheam MM, Biswas SK, Ahmed K et al (2021) Mycoremediation of reactive red H37B dye by *Aspergillus salinarus* isolated from textile effluents. Curr Res Microbial Sci 2:100056

Shekh AY, Shrivastava P, Krishnamurthi K, Mudliar SN, Devi SS, Kanade GS et al (2016) Stress enhances poly-unsaturation rich lipid accumulation in *Chlorella* sp. and *Chlamydomonas* sp. Biomass Bioenergy 84:59–66. https://doi.org/10.1016/j.biombioe.2015.11.013

Singh D, Goswami RK, Agrawal K et al (2022) Bioinspired remediation of wastewater: a contemporary approach for environmental cleanup. Curr Res Green Sustain Chem 5:100261

Srivastava A, Srivastava ON, Talapatra S, Vajtai R, Ajayan PM (2004) Carbon nanotube filters. Nat Mater 3:610–614

Stenholm Å, Hedel M, Arvidsson T, Pettersson C (2018) Identification of leachables from *Trametes versicolor* in biodegradation experiments. Trends Green Chem 4:1–4

Su Y (2021) Revisiting carbon, nitrogen and phosphorus metabolisms in microalgae for wastewater treatment. Sci Total Environ 768:144590. https://doi.org/10.1016/j.scitotenv.2020.144590

Su Y, Mennerich A, Urban B (2012) Synergistic cooperation between wastewater-born algae and activated sludge for wastewater treatment: influence of algae and sludge inoculation ratios. Bioresour Technol 105:67–73

Tong RM, Beck MB, Latten A (1980) Fuzzy control of the activated sludge wastewater treatment process. Automatica 16:675–701

Touliabah HE-S, El-Sheekh MM, Ismail MM et al (2022) A review of microalgae-and cyanobacteria-based biodegradation of organic pollutants. Molecules 27:1141

Ugwu CU, Aoyagi H, Uchiyama H (2007) Influence of irradiance, dissolved oxygen concentration, and temperature on the growth of *Chlorella sorokiniana*. Photosynthetica 45:309–311. https://doi.org/10.1007/s11099-007-0052-y

Vanessa de JG, Cristina MM, Almeida, AR, Elisabete, F, Maria JB, Vitor VC (2015) Occurrence of pharmaceuticals in a water supply system and related human health risk assessment. Water Res 72:199–208

Varjani S, Joshi R, Srivastava VK et al (2020) Treatment of wastewater from petroleum industry: current practices and perspectives. Environ Sci Pollut Res 80:1–9

Vélez E, Campillo GE, Morales G, Hincapié C, Osorio J, Arnache O et al (2016) Mercury removal in wastewater by iron oxide nanoparticles. J Phys: Conf Ser 687:012050

Vergili I (2013) Application of nanofiltration for the removal of carbamazepine, diclofenac and ibuprofen from drinking water sources. J Environ Manage 127:177–187

Vogelmann ES, Awe GO, Prevedello J (2016) Selection of plant species used in wastewater treatment. In: Wastewater treatment and reuse for metropolitan regions and small cities in developing countries. Publisher, pp 1–10

Von-Sperling M, Chernicharo CAL (2005) Biological wastewater treatment in warm climate regions, 1st edn. IWA Publishing, p 810

Wang K, Cai J, Xie S, Feng J, Wang T (2015) Phytoremediation of phenol using Polygonum orientale and its antioxidative response. Arch Environ Prot 41(3):39–46

Watkinson AJ, Murby EJ, Costanzo SD (2007) Removal of antibiotics in conventional and advanced wastewater treatment: implications for environmental discharge and wastewater recycling. Water Res 41(18):4164–4176

World Health Organization (2011a) Guidelines for drinking water quality. WHO, Geneva, Switzerland

World Health Organization (2011b) Pharmaceuticals in drinking-water. WHO Press, Geneva

Yewalkar SN, Dhumal KN, Sainis JK (2007) Chromium (VI)-reducing. Isolated from disposal sites of paper-pulp and electroplating industry. J Appl Phycol 19(5):459–465

Yuansong W, Renze RVH, Arjan RB, Dick HE, Yaobo F (2003) Minimization of excess sludge production for biological wastewater treatment. Water Res 27:4453–4467. https://doi.org/10.1016/S0043-1354(03)00441-X

Zahara A, Dahms H-U, Mika S, Eduardo HE, Roberto P (2020) Effectiveness of wastewater treatment systems in removing microbial agents: a systematic review. Global Health 12:2–10. https://doi.org/10.1080/17441692.2016.1273370

Zamora S, Marín-Muñíz JL, Nakase-Rodríguez C, Fernández-Lambert G, Sandoval L (2019) Wastewater treatment by constructed wetland eco-technology: influence of mineral and plastic materials as filter media and tropicalornamental plants. Water 11:2344

Zhang B, Li W, Guo Y, Zhang Z, Shi W, Cui F, Lens PNL, Tay JH (2020) Microalgal-bacterial consortia: from interspecies interactions to biotechnological applications. Renew Sustain Energy Rev 118:109563

Zhou X, Wang G (2010) Nutrient concentration variations during *Oenantheja vanica* growth and decay in the ecological floating bed system. J Environ Sci (China) 22:1710–1717

Zhou GJ, Ying GG, Liu S et al (2014) Simultaneous removal of inorganic and organic compounds in wastewater by freshwater green microalgae. Environ Sci 6:2018–2027

Zhu L, Li Z, Ketola T (2011) Biomass accumulations and nutrient uptake of plants cultivated on artificial floating beds in China's rural area. Ecol Eng 37:1460–1466

Zimmels Y, Kirzhner F, Roitman S (2004) Use of naturally growing aquatic plants for wastewater purification. Water Environ Res 76:220

Further Reading

Richard RK, Omondi DO, Wairimu MA, Maingi MS (2016) Influence of rainfall intensity on faecal contamination in River Nyangores of Mara Basin, Kenya: an eco-health integrity perspective. Asian J Microbiol Biotechnol Environ Sci 18(2):281–289

de Jesus Gaffney V, Almeida CMM, Rodrigues A, Ferreira E, Benoliel MJ, Cardoso VV (2015) Occurrence of pharmaceuticals in a water supply system and related human health risk assessment. Water Res 72:199–208

Chapter 4
Industrial Applications and Bioinformatics

Abstract Industrialization is straining global ecosystems, causing pollution and economic challenges. Microorganisms play a crucial role in environmental, economic, and social contexts, aiding in waste treatment and facilitating ecological transformations. Bioremediation, a popular cleaning technique, uses microorganisms to remove toxic waste from contaminated areas. Bioinformatics tools and biological databases help understand degradation processes, promoting sustainable remediation techniques and enhancing public health.

Keywords Wastewater · Microorganisms · Bioremediation · Food industry · Textile industry · Pharmaceutical industry · Petrochemical industry · Explosive industry · Distillery industry · Bioinformatics

4.1 Role of Microbes in Waste Management in Multiple Industrial Applications

The global ecological environment and terrestrial ecosystems are presently undergoing considerable strain due to human activities. The rapid and widespread expansion of industries has led to numerous milestones in growth, enhancing the efficiency and advancement of our lives. Economic growth around the world has been closely linked to industrialization, fundamentally transforming every sector of our existence. However, this progress comes with a serious downside, as industrial pollution represents a major challenge for modern society (Kaushik et al. 2012). The discharge from industrial activities includes harmful contaminants, both organic and inorganic that severely taint water bodies and the surrounding soil, thereby posing a detrimental effect on all living organisms (Chandra et al. 2015; Maszenan et al. 2011).

Microorganisms hold significant importance in environmental, economic, and social contexts. For centuries, they have been harnessed for producing several products, including enzymes, probiotics, and biofuels such as hydrogen gas and bioethanol. Presently, these microbes are crucial in industrial applications for the remediation of toxic waste. They are essential in breaking down of organic substances present in wastewater, serving as main contributors to the regulation of natural processes within ecosystems. As universal catalysts, they facilitate essential ecological transformations.

Following discussion focuses on the application of microorganisms in waste treatment across diverse sectors.

4.1.1 Pharmaceutical Industry

India is currently a major player in the pharmaceutical industry, coming in at number four in volume and thirteenth in value. Because it produces life-saving drugs, this business is regarded as the first science-based sector in India and is essential to contemporary civilization. However, because their metabolites negatively affect aquatic and terrestrial ecosystems, pharmaceutical chemical derivatives are important pollutants. Pharmaceutical wastewater is mostly made up of different medications and chemicals that can be difficult to manage, such as hormones, antibiotics, plant steroids, antidepressants, analgesics, lipid regulators, diuretics, and anti-inflammatory drugs.

This wastewater may include antibiotics such as triclocarban (TCC), erythromycin (ERY), sulfamethazine (SMI), naproxen (NAP), ibuprofen (IBU), gemfibrozil (GEM), 4-aminophenol, diphenhydramine (DPA), fluoxetine (FLU), trimethoprim (TMP), and diclofenac. Furthermore, there are hormones such as estriol (E3), ethinylestradiol (EE2), progesterone, 3,5,14(10)-tetrene, testosterone, estrogen, and 17b-trihydroxyestra-1 (Suresh and Abraham 2018). Additionally, the production of pharmaceuticals adds to the pollution problem.

Various physiochemical techniques can be employed to expedite and promote biodegradation of pharmaceutical waste, such as Fenton reactions, ultrasonic irradiation, and sophisticated oxidation techniques (including perozonation, TiO_2, oxidation, ozonation, and photocatalysis).

These techniques work well to remove a large amount of colloidal organic molecules and suspended materials, however they are ineffective when it comes to refractory compounds. Only microbial processes can lead to the mineralization of these stable and resistant organic molecules. These kinds of waste have historically been treated by bacteria (Table 4.1) (Rani et al. 2019). Both aerobic and anaerobic microbial processes may be utilized, and a variety of methods are employed, such as activated sludge, anaerobic film reactors, membrane batch reactors, anaerobic sludge reactors, and anaerobic filters.

The features of the wastewater itself have a major impact on how well biological treatments for wastewater work. However, in order to improve biodegradation and mineralization through biological treatments, a number of physical parameters must be optimized. These include organic load, dissolved oxygen levels, temperature, pH, hydraulic retention time, the presence of harmful and resistant chemicals, and the types of microorganisms. This technique holds great promise for the future since it can provide low-cost solutions with no waste byproducts. Microbial remediation offers

Table 4.1 Microbes existing within pharmaceutical effluent

Process		References
Fungal strain		
Bjerkandera adusta MUT 2295 Aspergillus sp., *Penicillium* sp.	Reduction in COD	Spina et al. (2012), Mohammad et al. (2006), Angayarkanni et al. (2003)
Bacterial strains		
Enterobacter		Ghosh et al. (2004)
Streptomonas Aeromonas Klebsiella	Phenol and complex organic compounds, Resorcinol	Kavitha and Beebi (2003)
Comamonas Arthrobacter Rhodococcus	Phenol/cresol phenol and catechol	Duffner et al. (2000), Kumar et al. (2005), Agarry and Solomon (2008)
Ralstonia B. thermoglucosidasius A7, Pseudomonas putida MTCC 1194 P. Fluorescence Rhodobacter sphaeroids	COD removal	Madukasi et al. (2010)

Reproduced with permission from Rani et al. (2019)

better pharmaceutical wastewater management options and encourages sustainable resource use when paired with other cutting-edge physiochemical and sophisticated oxidation techniques.

4.1.2 Food Industry

The natural water systems are seriously threatened by the wastewater produced by the food sector (Joshi and Deepali 2012). This wastewater comes from different parts of the food industry and is full of both organic and inorganic materials that can lower the oxygen content of water bodies they are placed in and harm a variety of organisms that live in ecosystems (Spina et al. 2012). The manufacture of vegetable oil, meat and poultry processing, bacon manufacturing, dairy and milk products, sugar refining, liquor, and brewery operations are all included in the food processing industry. The majority of food processing plants produce wastewater that is high in organic compounds and use a lot of water.

Chemical oxygen demand (COD), biochemical oxygen demand (BOD), temperature, pH, total suspended solids, total dissolved solids (TDS), and odor are important quality criteria for wastewater produced in the food business (Garg and Chaudhary 2017).

As time has progressed, the unsuitable management of wastewater generated by food industry operations has emerged as a pressing issue raising concerns for both governmental bodies and the industry itself. In many instances, even when industrial effluents undergo pretreatment through conventional methods, they frequently fail to satisfy the criteria necessary for effective removal of toxic pollutants. Consequently, these discrepancies have resulted in a substantial volume of water pollutants being released, posing a major health concern for both human beings and animals (Emmanuel et al. 2013).

A comprehensive analysis of the specific pollutants produced by various food industries is essential prior to the implementation of any particular wastewater treatment process. It is also essential to have a thorough awareness of the organizational structure, industrial production processes, and overall environmental effects of pollution. Wastewater treatment can be accomplished through various methods, which include physical, chemical, and biological methods, or a combination thereof (Emmanuel et al. 2013). Microorganisms such as algae, fungi, and bacteria play a crucial role in the biological processes involved in wastewater treatment. The organic waste present in industrial effluents serves as a nutrient source for many bacteria, as they require elements like phosphorus, nitrogen, and other trace nutrients for their growth.

McIlory et al. (2015) conducted study across 20 wastewater treatment facilities. This study revealed that the predominant bacterial genera included *Microthrix, Tetrasphaera, Hyphomicrobium, Rhodobacter, Rhodoferax, Candidatus*, and *Trichococcus*. Furthermore, Zhang et al. (2012) recognized additional microorganisms, including *Zoogloea , Caldilinea, Dechloromonas, Prosthecobacter*, and *Tricoccus*, that were consistently found across all wastewater treatment facilities. Nevertheless, the diversity of species present in the biomass differed according to their geographical settings.

Frequently occurring microbial entities in activated sludge from a total of fourteen plants designed for the processing of wastewater included *Anaerolineales, Sphingobacteriales, Planctomycetales, Rhizobiales, Burkholderiales, Xanthomonadales, Anaerolineales, Rhodocyclales*, and *Clostridiales*.

4.1.3   Textile Industry

The textile sector is characterized by its significant water consumption, necessitating substantial amounts of water for the purpose of applying chemicals and washing finished textiles (Ntuli et al. 2009). This high demand for water results in the production of considerable volumes of effluent with diverse compositions during dyeing and finishing operations. The chemical burden identified in the effluent produced through the process of textile manufacturing has become a pressing issue in contemporary discussions. This load consists of diverse byproducts, auxiliary chemicals, and leftover dyes, which contribute to the water's high levels of coloration, alkalinity, and subsequently elevate its temperature, BOD, and suspended solids (U.S. EPA 1997; Mohan et al. 2007).

The release of wastewater from textile manufacturing processes fails to meet established regulatory standards, leading to widespread pollution of various environmental resources, such as surface soils and aquatic ecosystems, due to its high levels of dyes and other hazardous chemicals (Rani et al. 2019). The regulation of wastewater arising from textile production facilities can be achieved through various approaches, including chemical, physical, and biological treatments.

The biological approach to treating wastewater has consistently proven to be a highly cost-effective option. This approach utilizes a diverse range of microbes, including fungi, yeasts, bacteria, and algae, which are able to accumulate and decompose various pollutants. These microorganisms utilize the components of the effluent as their nutrient source, leading to an increase in their biomass. During this process, they decompose complex compounds into simpler and less harmful substances, thereby reducing BOD and COD.

In the textile industry, the predominant methods utilized in the process of wastewater treatment incorporate microbial biomass to perform adsorption, decolorization, and degradation processes (Fu and Viraraghavan 2001). The activated sludge process is a widely adopted biological treatment approach. Initially, the wastewater undergoes cooling and screening, followed by homogenization in a tank to create a uniform effluent with varying characteristics. Subsequently, neutralization is performed using acids such as sulfuric acid or hydrochloric acid to establish an acidic environment (pH 6.5–8.5) conducive to bacterial activity. The effluent is then routed to a biological oxidation tank after this step, where microbes further break it down into smaller molecular components. It is essential to maintain appropriate levels of dissolved oxygen (DO), pH, temperature, and nutrient availability to ensure optimal microbial performance.

The fungi that have been extensively researched for the biological management of industrial wastewater include *Rhizopus oryzae*, *Aspergillus foetidus*, *Penicillium geastrivous*, and *Umbelopsis isabellina* among others (Yang et al. 2003). These fungi are significant contributors to the processes of dye decolorization and adsorption. Utilizing fungi for decolorization presents a promising alternative to traditional treatment methods. Among them, *Irpex lacteus* and *Pleurotus ostreatus* are more effective than *Aspergillus ustus* and *Dactylospora haliotrepha* at breaking down dyes in different categories (Marimuthu et al. 2013).

A number of aerobic bacteria are also found to decolorize colors. These include *Bacillus* sp., *Klebsiella* sp., *Pseudomonas fluorescens*, *Enterococcus sp*sp., and *Shewanella putrefaciens*. The biodegradation of phenol is notably facilitated by various bacteria, including *Pseudomonas* sp., *Achromobacter*sp., and *Acinetobacter* sp., in addition to fungal species like *Phanerochaete chrysosporium, Coriolus versicolor, Fusarium*sp., *Ralstonia* sp., and *Streptomyces* sp. Furthermore, for degrading chromium, both fungi and bacteria, including *Bacillus* sp., *Arthrobacter* sp., *Acinetobacter*, *Brevibacterium casei*, *E. coli*, and *P. aeruginosa*, have shown considerable effectiveness (Das and Mishra 2010; Srivastava and Thakur 2007).

4.1.4   *Explosive Industry*

The present military environment has experienced a prominent increase in the creation of explosives employed by defense forces for an array of functions. Significant energetic explosives encompass a variety of substances like nitrocellulose (NC), hexahydro-1,3,5-trinitro-1,3,5-triazine, known as RDX, 2,4,6-trinitrotoluene and octahydro-1,3,5,7-tetranitro-1,3,5,7-tetrazocine, generally referred to as HMX, among others. The processes of manufacturing, packaging, transporting, and incinerating these substances contaminate water and surrounding soil. The byproducts of these explosives are resistant to environmental biodegradation, placing them on the priority list of USEPA owing to their harmful and mutagenic characteristics. TNT, a commonly utilized explosive in World War II, has been linked to numerous health concerns including cataracts, aplastic anemia, muscular pain, and cyanosis (Yang et al. 2008).

The role of microorganisms is significant in the bioremediation efforts aimed at restoring land and treating wastewater contaminated with explosives. Research has shown the ability of fungi and bacteria to degrade and transform these explosives. A list of some microorganisms effective in degrading explosives is provided in Table 4.2 (Rani et al. 2019). Investigations into microbial profiling and metagenomics concerning water and soil tainted by energetic compounds have substantiated the involvement of microbes in the degradation process of explosive materials. For instance, microbial profiling of groundwater at a specific contaminated site showed a total of 1605 operational taxonomic units of which ninety-six were found to be bacteria with the most common being *Rhodococcus*. The contaminated groundwater was predominantly composed of Proteobacteria and Actinobacteria (Wang et al. 2017).

Various fungi have been identified in sites contaminated with TNT, including *Trichoderma*, *Alternaria*, and *Aspergillus* (Bennett et al. 1995). *P. chrysosporium*

Table 4.2 Microbes used for breaking down various explosives

Name of explosive	Degrading microorganism	References
CL-20 TNT	*Clostridium* sp., *Pseudomonas*sp.*MS-4*	Panikov et al. (2007), Crocker et al. (2006), Sangwan et al. (2015), Boopathy and Manning (1998)
RDX, HMX 4-nitrophenol 2,4-dinitrophenol TNT, DNT, Nitrobenzene	*Acinetobacter noscomialis, Sulfate-Reducing and Methanogenic Consortia Clostridium* sp. *Bacillus sphaericus JS905 Rhodococcus* sp. *Desulfovibrio* sp. *Strain B, Pseudomonas* sp. *JX165*	Crocker et al. (2006), Kadiyala and Spain (1998), Takeo et al. (2003), Boopathy and Kulpa (1993)

Reproduced with permission from Rani et al. (2019)

has demonstrated the ability to degrade CL-20 (Hexanitrohexaazaisowurtzitane) in an aqueous environment (Karakaya et al. 2009). The degradation is affected by the availability of nitrogen and carbon sources. Research indicates that the growth of *P. chrysosporium* is not impacted at levels reaching 500 mg/L.

Sangwan et al. (2015) reported that microbial degradation and transformation of explosives are more effectively carried out by native flora. *P. aeruginosa* has been shown to degrade TNT into various metabolites, including 2,4-dinitrotoluene and 4-aminodinitrotoluene. This degradation is dependent on both concentration and duration, as well as the growth of the cellular biomass contained within the medium (Mercimek et al. 2015). Furthermore, the bacterial strain *Acenitobacter noscomialis* was found to degrade TNT in an aqueous phase, converting it into diaminonitrotoluenes, aminodinitrotoluene, hydroxylaminodinitrotoluene.

Acenitobacter noscomialis is capable of functioning across a range of TNT concentrations in aqueous environments. Abiotic factors exert a minimal influence on the process of explosive degradation. The breakdown of RDX and TNT in saline water is predominantly independent of salinity levels and UV radiation, although a minor impact of ultraviolet radiation on the deterioration of TNT was observed. Despite this, abiotic factors may have an impact on the photolytic reaction of TNT and the volatilization phenomenon of RDX to a limited degree (Sisco et al. 2015).

The breakdown of energetic materials is contingent upon various parameters, including the presence of oxygen. There is a scarcity of reports concerning the aerobic degradation of specific explosives, including HMX and RDX. Species of bacteria, including *Gordonia* and *Williamsia*, have been recognized for their ability to aerobically biodegrade RDX, using it as their exclusive nitrogen and carbon source, which leads to the formation of formaldehyde and nitrite as byproducts (Fida et al. 2014).

Indigenous soil bacteria play a crucial role in the degradation of explosives like 2,4,6-trinitrotoluene and PETN (Pentaerythritol tetranitrate); however, their rate and efficiency in this process are somewhat constrained (Amin et al. 2017). The biodegradability of these substances can be enhanced through certain soil amendments, including the introduction of biosurfactants like mono-rhamnolipid; this enhances the bioavailability of explosives for bacterial absorption and promotes their permeability through the cell membrane. This indicates that native bacterial species have the capability to use TNT and PETN as their exclusive sources of energy. The degradation of explosives is also influenced by the presence of specific metabolizing enzymes within the bacterial cells. XenA and XenB reductases, members of the flavoprotein oxidoreductase family, contribute to the degradation of various energetic substances both in anaerobic and aerobic environments. The success of this degradation process is contingent upon the particular explosive employed and the presence of oxygen. Research has shown that in *Pseudomonas* species, RDX is transformed into formaldehyde via methylene dintramine in low-oxygen conditions, without the direct reduction of the nitro groups in RDX (Fuller et al. 2009).

A study was performed using genetically modified *P. fluorescens* that expresses the XplA gene to evaluate RDX degradation within the rhizosphere. The research revealed that the function of the XplA gene requires an additional component, namely α-aminolevulinic acid, which acts as a precursor for heme. This study was initiated due to the lack of identified RDX-degrading bacteria that can inhabit the rhizosphere, despite the presence of significant microbial activity (Lorenz et al. 2013).

Microbial techniques have demonstrated considerable success in the process of restoring polluted water and soil to a safe and usable condition. Despite extensive research over the past two decades on the isolation of microbes and the biodegradation of explosive compounds, a comprehensive technology for complete mineralization remains unavailable. Consequently, it is essential to develop innovative techniques such as composting and biostimulation, as well as to investigate microbial interactions with explosives, to facilitate the effective removal of these contaminants from the environment.

4.1.5   Distillery Industry

Distilleries are integral to the agricultural industry that rely on various agricultural products, such as wheat, molasses, rice, barley, sugar cane juice, sugar beet molasses, corn, and cassava (Kawa-Rygielska et al. 2007; Wilkie et al. 2000). The production process results in significant volumes of wastewater, commonly referred to as spent wash. For each liter of alcohol generated, it is estimated that 10–15 L of wastewater are also produced. This waste arises during multiple stages of the production process, such as distillation, cooling in condensers and fermenters, fermentation, and washing. The features of this effluent are influenced by the type of feedstock utilized in the distillery. It typically exhibits a dark coloration, lower pH, elevated temperature, and a higher concentration of both organic and inorganic matter.

The wastewater produced by distillery activities presents a significant environmental risk if not sufficiently administered. It releases harmful substances into both the soil and aquatic ecosystems. The presence of dark brown hues in this wastewater diminishes the availability of light in aquatic environments, which adversely impacts both flora and fauna. The infiltration of this substance into the soil has the potential to reduce soil alkalinity while simultaneously elevating the concentrations of heavy metals, including mercury, nickel, silver, copper, iron, and cadmium. This hue is predominantly a result of the presence of melanoidin, a polymer generated through the Maillard reaction, which is notably resistant to environmental degradation and poses toxicity to various microorganisms because of its antioxidant characteristics. This can lead to decreased soil fertility and inhibited microbial activity.

The color of spent wash is influenced by various additional compounds, including phenolic substances such as humic acid, tannic acid, proteins, caramels produced from sugars that have been overheated, and furfurals formed as a result of acid hydrolysis (Kort 1979). It is crucial that distillery effluent undergoes treatment before being released into the environment. A range of physicochemical treatment methods is generally utilized, tailored to the specific chemical composition of the spent wash. The techniques involved include screening, coagulation, adsorption, sedimentation, pH correction, flotation, electrolysis, and ultrafiltration. Nonetheless, these techniques frequently encounter challenges related to costs, chemical consumption, significant sludge generation, and the production of additional byproducts.

There has been a significant rise in worldwide research focus on the application of microbial techniques for wastewater treatment. This strategy is characterized by its environmental sustainability and economic practicality. Numerous organisms are capable of breaking down and decolorizing wastewater. These microorganisms are crucial to the treatment process, functioning through various mechanisms including enzymatic degradation, adsorption, and absorption. The enzymatic breakdown performed by these microorganisms facilitates the decolorization and removal of melanoidin, as they utilize the pigment for carbon and nitrogen. Furthermore, the process of flocculation is improved by extracellular compounds and the adsorption onto both viable and non-viable biomass (Pant and Adholeya 2010; Chandra et al. 2009).

Numerous enzymes are produced by microorganisms, both intracellularly and extracellularly. These enzymes include laccases, sugar oxidases, lignin peroxidases, and manganese peroxidases all of which contribute to the degradation of melanoidins (Couto et al. 2005; Freitas et al. 2009). In the process of degradation, melanoidins are adsorbed onto microbial cell structures, followed by their incorporation and breakdown by intracellular enzymes. This reaction entails the utilization of active oxygen along with sugar molecules present in the mixture. The enzymatic oxidation process primarily generates active oxygen species, with hydrogen peroxide being the predominant component which is produced in the presence of various sugars, including lactose, glucose, maltose, sorbose, xylose, and sucrose (Pant and Adholeya 2007; Watanabe et al. 1982).

White rot fungi feature an advanced extracellular enzymatic system, enabling them to break down poly-aromatic compounds, lignin, and melanoidins, even in nutrient-poor environments (Benito et al. 1997). *Acetobacter acetii, Alicaligens* sp. , *Aspergillus niger, Bacillus* sp. , *Chlorella* sp. , *Coriolus hirsutus Flavodon flavus, Fusarium*sp. ., *Geotrichum candidum, Lactobacillus hilgardii WNS, Lyngbya, Nostoc muscorum Oscillatoria boryna, P. chrysosporium, P. fluorescenes, Rhizoctonia sp. D-90, Synechocystis, Trametes versicolor* are the most often researched bacteria, fungi, and algae in relation to distillery effluent (Rani et al. 2019).

Microalgae present considerable advantages for bioremediation, as they can utilize contaminants such as nitrate, phosphate, and ammonium, as nutrients and demonstrate a remarkable ability to flourish in more demanding environments than bacteria and fungi.

Moreover, microalgae can be harvested for a range of valuable products, including ethanol, methane, livestock feed, and organic fertilizer. Several cyanobacterial species have been identified as capable of degrading and decolorizing melanoidin.

4.1.6 *Petrochemical Industry*

The petrochemical sector consists of chemicals obtained from petroleum and natural gas, which include synthetic rubber, yarn, plastics, detergents, polymers, and various processing industries.

This sector emerged in the 1970s, but it was not until the 1990s that significant growth was observed within the Indian context. Currently, the petrochemical industry has integrated itself into nearly all dimensions of daily life, affecting sectors including food and water security, healthcare, societal and physical infrastructure, domestic goods, building materials, vehicles, telecommunications, electronic devices, farming, and gardening. The widespread use of petrochemical products has resulted in pollution affecting various aspects of life, significantly harming environmental resources, especially in areas designated for petrochemical manufacturing. This has resulted in the pollution of air, water, and oil. The effluent produced by a standard petrochemical refinery comprises a diverse array of organic compounds present in elevated concentrations which are resistant to environmental degradation. The toxic substances produced include aromatic and aliphatic hydrocarbons, cyanide, octanols, phenols, and formaldehyde (Yeruva et al. 2015). Current physicochemical treatment techniques, such as membrane extraction, photocatalytic oxidation, and flocculation-coagulation, encounter challenges related to both operational and capital expenditures, in addition to generating waste (Verma et al. 2010; Madaeni and Eslamifard 2010). The harmful chemicals produced often fail to meet wastewater disposal standards and pose significant risks as immune toxicants and carcinogens to living organisms.

In order to realize long-term objectives related to price, risk management, effectiveness, environmental sustainability, simplicity, bioremediation, which involves the use of microbes for wastewater treatment, stands as the sole solution. A diverse array of microbes, including yeasts, fungi, bacteria, algae, and protozoa, possess the capability to degrade hydrocarbons (Table 4.3) (Rani et al. 2019).

These organisms are recognized for their ability to restore polluted environments (Thangaraj et al. 2007). Typically categorized as chemoorganotrophs, these microbes use organic substances as suppliers of carbon and donors of electrons to facilitate ATP synthesis. Certain microbes exploit the hydrophobic characteristics of hydrocarbons, facilitating degradation through the synthesis of biosurfactants (Das and Mukherjee 2007).

Specific mechanisms have been devised by microbes, especially bacteria, to thoroughly decompose hydrocarbons that are not soluble in water. This procedure involves an adhesion mechanism along with the synthesis of extracellular polysaccharides; these agents act as emulsifiers to increase the effectiveness of contact (Hisatsuka et al. 1971).

The degradation of petrochemicals can occur through both aerobic and anaerobic processes. A standard aerobic process encompasses enzyme oxygenases that facilitate the incorporation of molecular oxygen into the reduced substrate, leading to the first emergence of alcohols from aliphatic hydrocarbons. Following their

Table 4.3 Microbes involved in the treatment of petrochemical wastewater

Phenolic compounds, PAHs, and oils	Distillation unit, Hydro-Cracking, catalytic Cracking, Spent caustic, vis breaker, Ballast water, lube oil, utility	*Sphingobacterium*	Khongkhaem et al. (2011), Ho et al. (2012), Chaudhary and Kim (2016), Qu et al. (2015), Gomez-Acata et al. (2016), Cao and Wang (2004), Mohanty and Mukherji (2008), Thangaraj et al. (2007), Margesin et al. (2005), Thangaraj et al. (2007), Shokrollahzadeh et al. (2008), Moller et al. (1996), Shim et al. (2005), Pandey et al. (2007)
Phenolic compounds and PAHs, COD		*Pseudomonas Comamonas Rudaea Brevundimonas Acinetobacter calcoaceticus Alcaligenes odorans*	
n-alkanes fraction of hydrocarbons			
Phenol, dibenzothiophene and dibenzofuran		*Ralstonia Burkholderia cepacia Caulobacter* sp. *Flavobacterium* sp. *Janibacter terrae*	
Fluorene biphenyls, 4-chlorophenol Alkanes, aromatic, polycyclic hydrocarbons H$_2$S, Ammonia Odor		*Sphingomonas paucimobilis Rhodococcus rhodochrous*	
		Acidovorax, Flavobacterium, Cytophaga Bacillus sphaericus and Pseudomonas putida	

Reproduced with permission from Rani et al. (2019)

initial formation, these alcohols are acted upon by dehydrogenases, leading to their oxidation into carboxylic acids, which subsequently participate in the process of β-oxidation. Hydroxylation of the rings in aromatic and polycyclic aromatic hydrocarbons occurs through the action of mono- or dioxygenases, culminating in the formation of diols. Following this process, ring cleavage occurs, which is subsequently metabolized by prokaryotes and eukaryotes. Oil biodegradation may happen in anaerobic environments, supported by bacteria reducing sulfates (Holba et al. 1996; Cerniglia and Yang 1984). Studies indicate that sulfogenic bacteria and methanogenes utilize oxidative mechanisms to break down various hydrocarbons, such as alkanes, toluene, and benzene, especially in environments characterized by strict anaerobiosis and denitrification (Grabic-Galic and Vogel 1987; Ward and Brock 1978).

4.2   Integrating Bioinformatics into Bioremediation Practices

Bioremediation represents an important method in managing waste effectively aimed at the removal of waste from polluted sites. This approach emphasizes the function of organisms in either consuming or mitigating contaminants existing in the surroundings (Giri et al. 2021). Bioremediation seeks to elucidate the degradation mechanisms that specific organisms use for particular pollutants by leveraging data from a range of biological databases, such as those that detail chemical structures, RNA and protein expression, catalytic enzymes, organic substances, pathways involved in microbiological deterioration, and systematic comparison of genomic data from multiple species. This study is made easier by a variety of bioinformatics tools, which contributes to the advancement of environmentally friendly cleanup solutions (Khanna and Kumar 2022; Bala et al. 2022; Kumari and Kumar 2021; Nigam and Sinha 2023; Fulekar and Sharma 2008; Fulekar 2009; de Lorenzo 2008; Arora and Bae 2014; Arora and Shi 2010; Arora et al. 2022; Khan 2018).

4.2.1   Bioinformatics

Bioinformatics is a branch of science that combines biology and information technology to analyze large biological data sets. This encompasses the creation of statistical instruments and algorithms aimed at analyzing and identifying relationships among biological data sets, including macromolecular sequences, structures, expression profiles, and biochemical pathways. The discipline of bioinformatics has recently expanded its scope to include bioremediation. Proteomics, biological databases, data mining, molecular phylogenetics, microarray informatics, genomics, and systems biology are important subfields in bioinformatics. The tools developed in bioinformatics are essential to the bioremediation of hazardous waste and the advancement of environmental cleanup technologies. Proteomics, a key post genomic feature, is essential for understanding the actions of proteins in bioremediation.

The variety of applications for bioremediation is limited by the lack of understanding of the variables affecting microbial growth and metabolism with the potential for bioremediation. These bacteria have been thoroughly profiled, and bioinformatics approaches have been used to clarify their mineralization processes and mechanisms. Additionally, the analysis of methodologies and technologies related to bioremediation heavily relies upon the application of proteomic methods, like microarrays, two-dimensional polyacrylamide gel electrophoresis, and mass spectrometry. Studies have shown that these techniques greatly improve the examination of the structure of microbial proteins that can effectively break down pollutants.

The advancement in the structural characterization of such proteins has been substantial, and this research area integrates aspects of computer science and biology, utilizing computational tools to handle, store, and retrieve genomic data pertaining to proteins, RNA, and DNA (Zheng et al. 2018; Vega-Páez et al. 2019; Yergeau et al. 2012).

4.2.2 Omics-Driven Solutions for Bioremediation Efforts

Bioremediation research can significantly benefit from the utilization of omics technologies, which encompass genomics, transcriptomics, metabolomics, and proteomics. These technologies facilitate the linking of DNA sequences to the levels of proteins, metabolites, and mRNA thereby improving the assessment of bioremediation processes occurring in situ (Sar and Islam 2012; Villegas-Plazas et al. 2019).

4.2.3 Genomics

The study of bioremediation bacteria has given rise to a new specialty in genetics. The foundation of this tactic is the capacity of microorganisms to completely examine their genetic data inside of cells. In bioremediation, a wide variety of microorganisms are employed (Jaiswal et al. 2019). Genomic methods like Polymerase Chain Reaction (PCR), isotope distribution analysis, DNA hybridization, metabolic footprinting, molecular connectivity, and genetic engineering are utilized in order to comprehend the biodegradation process better. Numerous PCR-based methods are available for genotypic fingerprinting (Hakeem et al. 2020):

- Random amplified polymorphic DNA analysis (RAPD);
- Automated ribosomal intergenic spacer analysis (ARISA);
- Terminal-restriction fragment length polymorphism (T-RFLP);
- Amplified ribosomal DNA restriction analysis (ARDRA);
- Amplified fragment length polymorphisms (AFLP);
- Single strand conformation polymorphism (SSCP);
- Length heterogeneity.

RAPD can be used to evaluate bacterial species that are fundamentally connected, build structural models with functionality and generation of genetic fingerprints when examining soil microbial communities. LH-PCR can be utilized for identifying the inherent length differences observed among distinct SSU rRNA genes in microbial population. T-RFLP allows for the concurrent profiling of several taxonomic groupings of microorganisms (Rodríguez et al. 2022; Gupta et al. 2020).

A range of molecular methods, including genetic fingerprinting, FISH, microradiography, quantitative PCR, and stable isotope probing, can be utilized in studies investigating the interactions between soil bacteria and environmental variables. The presence and quantity of operational and taxonomic gene markers in the soil can be ascertained by a quantitative PCR study of the soil's microbial communities. DNA analysis approaches utilize products from amplified PCR as a foundation for prompt assessment of particular genes for molecular biomarkers (Yunusa and Umar 2021). By comparing the fingerprints from various samples, cluster-assisted analysis can aid in clarifying the relationship between distinct microbiological groups.

4.2.4   Transcriptomics and Metatranscriptomics

The transcriptome is a vital connection between the genome, interactome, proteome, and cellular phenotype. It consists of the group of genes which are actively expressed under particular situations and at specific periods. In order for organisms to adapt to changes in their environment and increase their chances of survival, gene expression regulation is essential. Transcriptomics offers a holistic perspective on this process across the human genome. Among the various methodologies in transcriptomics, DNA microarray analysis stands out as particularly effective for assessing mRNA expression levels. The transcriptomic analysis process initiates using complete mRNA isolation and enrichment, followed by the synthesis of complementary DNA (cDNA), culminating in the structuring of the cDNA transcriptome. The application of a DNA microarray in transcriptomics facilitates the analysis of mRNA expression across nearly all genes present in an organism. The examination of transcriptional mRNA profiles, referred to as metatranscriptomics or transcriptomics, is crucial for acquiring a functional comprehension of the dynamics present within environmental microbial communities. The relationships among microbes and their associated metabolic pathways can be clarified through the use of metagenomics, genome binning, and metatranscriptomics, particularly in the realm of biodegradation (Sharma et al. 2022; Singh et al. 2022; Haque et al. 2022; Lawrence et al. 2013; Zhang et al. 2010).

4.2.5   Proteomics and Metabolomics

Proteomics is the complete examination of every protein in a cell at a particular place and moment. This differs from metabolomics, which investigates the complete range of metabolites that an organism produces over a defined time frame or under specific conditions. Through the application of proteomics, researchers have effectively studied protein abundance, variations in composition, and key proteins linked to microbiological organisms.

When evaluated alongside genomics, microbial communities' functional study shows more potential and utility. Metabolomics can be approached in two primary ways to investigate biological systems. The first approach does not require preliminary awareness of the metabolic pathways present in the biological system, allowing for the identification and recovery of numerous metabolites from samples, thus

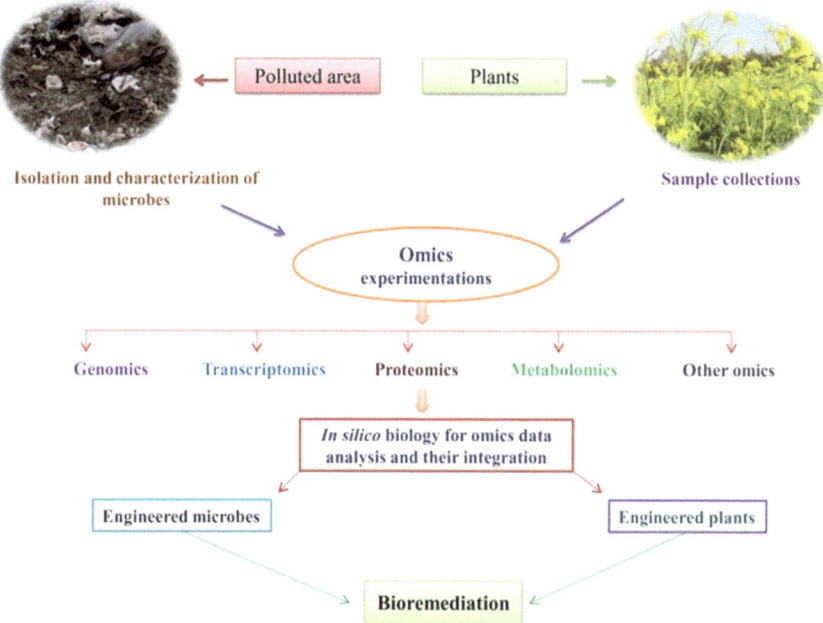

Fig. 4.1 Mechanism of bioremediation via integration and analysis of multi-omics data. Reproduced with permission from Verma et al. (2021)

producing extensive data that elucidates the relationships among various samples within metabolic pathways.

Alternatively, a targeted approach can be employed to focus on specific metabolic pathways or metabolites identified through the analysis of existing studies. A variety of tools, such as metabolite profiling, footprinting, and target analysis, are utilized in microbial metabolomics for the identification and quantification of the diverse cellular byproducts generated by living organisms. Insights gained from the metabolome and proteome will be valuable for initiatives focused on cell-free bioremediation (Tripathi et al. 2018; Gaur et al. 2022; Bharagava et al. 2019; Sanghvi et al. 2020).

In Fig. 4.1, the mechanism underlying bioremediation is presented, highlighting the integration and analysis of multi-omics data (Verma et al. 2021).

References

Agarry SE, Solomon BO (2008) Kinetics of batch microbial degradation of phenols by indigenous *Pseudomonas fluorescence*. Int J Environ Sci Technol 5(2):223–232

Amin MM, Khanahmad H, Teimouri F, Sadani M, Karami MA, Rahimmanesh I (2017) Improvement of biodegradability of explosives using anaerobic-intrinsic bioaugmentation approach. Bul Chem Commun 49:735–741

Angayarkanni J, Palaniswamy M, Swaminathan K (2003) Biotreatment of distillery effluent using *Aspergillus niveus*. Bull Environ Contam Toxicol 70:268–277

Arora PK, Bae H (2014) Integration of bioinformatics to biodegradation. Biol Proced Online 16:8. https://doi.org/10.1186/1480-9222-16-8

Arora P, Shi W (2010) Tools of bioinformatics in biodegradation. Rev Environ Sci Biotechnol 2010(9):211–213. https://doi.org/10.1007/s11157-010-9211-x

Arora PK, Kumar A, Srivastava A, Garg SK, Singh VP (2022) Current bioinformatics tools for biodegradation of xenobiotic compounds. Front Environ Sci 10:980284. https://doi.org/10.3389/fenvs.2022.980284

Bala S, Garg D, Thirumalesh BV, Sharma M, Sridhar K, Inbaraj BS, Tripathi M (2022) Recent strategies for bioremediation of emerging pollutants: a review for a green and sustainable environment. Toxics 10(8):484. https://doi.org/10.3390/toxics10080484

Benito GG, Miranda MP, Santos DR (1997) Decolorization of wastewater from an alcoholic fermentation process with *Trametes versicolor*. Biores Technol 61:33–37

Bennett JW, Hollirah P, Waterhouse A, Horvath K (1995) Isolation of bacteria and fungi from TNT contaminated composts and preparation of 14C-ring labelled TNT. Int Biodeterior Biodegradation 35(4):421–430

Bharagava RN, Purchase D, Saxena G, Mulla SI (2019) Applications of metagenomics in microbial bioremediation of pollutants: from genomics to environmental cleanup. Microbial diversity in the genomic Era. Academic Press, Cambridge, MA, USA, pp 459–477

Boopathy R, Kulpa CF (1993) Nitroaromatic compounds serve as nitrogen source for *Desulfovibrio* sp. (B strain). Can J Microbiol 39(4):430–433

Boopathy R, Manning J (1998) A laboratory study of the bioremediation of 2,4,6-trinitrotoluene contaminated soil using aerobic anaerobic soil slurry reactor. Water Environ Res 70:80–86

Cao W, Wang Y (2004) Compound-specific hydrogen and carbon isotopic fractionations during artificial enhanced bioremediation of petroleum hydrocarbons. In: Proceedings of 227th ACS national meeting, Anaheim, CA, USA, March 28–April 1, GEOC-028. American Chemical Society, Washington D.C.

Cerniglia C, Yang S (1984) Stereoselective metabolism of anthracene and phenanthrene by the fungus *Cunninghamella elegans*. Appl Environ Microbiol 47(1):119–124

Chandra R, Bharagava RN, Rai V, Singh SK (2009) Characterization of sucrose-glutamic acid Maillard products (SGMPs) degrading bacteria and their metabolites. Biores Technol 100:6665–6668

Chandra R, Saxena G, Kumar V (2015) Phytoremediation of environmental pollutants: an eco-sustainable green technology to environmental management. In: Chandra R (ed) Advances in Biodegradation and bioremediation of industrial waste. CRC Press, Boca Raton, FL, pp 1–30

Chaudhary DK, Kim J (2016) *Sphingomonas naphthae* sp. nov., isolated from oil-contaminated soil. Int J Syst Evol Microbiol 66:4621–4627

Couto SR, Sanroman MA, Gubitz GM (2005) Influence of redox mediators and metal ions on synthetic acid dye decolorization by crude laccase from *Trametes hirsuta*. Chemosphere 58:417–422

Crocker FH, Indest KJ, Fredrickson HL (2006) Biodegradation of the cyclic nitramine explosives RDX, HMX, and CL-20. Appl Microbiol Biotechnol 73(2):274–290

Das AP, Mishra S (2010) Biodegradation of metallic carcinogen hexavalent chromium Cr(VI) by an indigenously isolated bacterial strain. Journal of Carcinogenesis 9:6

Das K, Mukherjee AK (2007) Crude petroleum-oil biodegradation efficiency of *Bacillus subtilis* and *Pseudomonas aeruginosa* strains isolated from a petroleum-oil contaminated soil from North-East India. Biores Technol 98:1339–1345

de Lorenzo V (2008) Systems biology approaches to bioremediation. Curr Opin Biotechnol 19:579–589

Duffner FM, Kirchner U, Bauer MP, Muller R (2000) Phenol/cresol degradation by the thermophilic *Bacillus thermoglucosidasius* A7: cloning and sequence analysis of five genes involved in the pathway. Gene 256:215–221

Emmanuel A, Jacob E, Liberty T (2013) Effluents characteristics of some selected food processing industries in Enugu and Anambra states of Nigeria. J Environ Earth Sci 3(9):46–53

Fida TT, Palamuru S, Pandey G, Spain JC (2014) Aerobic biodegradation of 2,4-dinitroanisole by *Nocardioides* sp. strain JS1661. Appl Environ Microbiol 80(24):7725–7731

Freitas AC, Ferreira F, Costa AM, Pereira R, Antunes SC (2009) Biological treatment of the effluent from a bleached kraft pulp mill using basidiomycete and zygomycete fungi. Sci Total Environ 407:3282–3289

Fu Y, Viraraghavan T (2001) Fungal decolorization of dye wastewater-review. Biores Technol 79:251–262

Fulekar MH (2009) Bioinformatics: applications in life and environmental sciences. Springer Science & Business Media, Germany

Fulekar MH, Sharma J (2008) Bioinformatics applied in bioremediation. Innovative Rom Food Biotechnol 3:28–36

Fuller ME, McClay K, Hawari J, Pauquet L, Malone TE, Fox BG, Steffan RJ (2009) Transformation of RDX and other energetic compounds by xenobiotic reductases XenA and XenB. Appl Microbiol Biotechnol 84(3):535–544

Garg S, Chaudhary S (2017) Treatment of wastewater of food industry by membrane bioreactors. Int Adv Res J Sci Eng Technol 4(6):153–156

Gaur VK, Gautam K, Sharma P, Gupta P, Dwivedi S, Srivastava JK, Varjani S, Ngo HH, Kim SH, Chang JS et al (2022) Sustainable strategies for combating hydrocarbon pollution: special emphasis on mobil oil bioremediation. Sci Total Environ 832:155083

Ghosh M, Verma SC, Mengoni A, Tripathi AK (2004) Enrichment and identification of bacteria capable of reducing chemical oxygen demand of anaerobically treated spent wash. J Appl Microbiol 6:241–278

Giri BS, Geed S, Vikrant K, Lee SS, Kim KH, Kailasa SK, Vithanage M, Chaturvedi P, Rai BN, Singh RS (2021) Progress in bioremediation of pesticide residues in the environment. Environ Eng Res 26:200446

Gomez-Acata S, Esquivel-Rios I, Perez-Sandoval MV, Navarro-Noya Y, Rojas-Valdez A, Thalasso F, Luna-Guido M, Dendooven L (2016) Bacterial community structure within an activated sludge reactor added with phenolic compounds. Appl Microbiol Biotechnol 101:3405–3414. https://doi.org/10.1007/s00253-016-8000-z

Grabic-Galic D, Vogel TM (1987) Transformation of toluene and benzene by mixed methanogenic culture. Appl Environ Microbiol 53(2):254–260

Gupta K, Biswas R, Sarkar A (2020) Advancement of omics: prospects for bioremediation of contaminated soils. In: Microbial bioremediation & biodegradation. Springer, Singapore, pp 113–142

Hakeem KR, Bhat RA, Qadri H (2020) Bioremediation and biotechnology. Springer, Cham, Switzerland

Haque S, Srivastava N, Pal DB, Alkhanani MF, Almalki AH, Areeshi MY, Naidu R, Gupta VK (2022) Functional micro-biome strategies for the bioremediation of petroleum-hydrocarbon and heavy metal contaminated soils: a review. Sci Total Environ 833:155222

Hisatsuka K, Nakahara T, Sano N, Yamada K (1971) Formation of rhamnolipid by *Pseudomonas aeruginosa* and its function in hydrocarbon fermentation. Agric Biol Chem 35:686

Ho YN, Mathew DC, Hsiao SC, Shih CH, Chien MF, Chiang HM, Huang CC (2012) Selection and application of endophytic bacterium *Achromobacter xylosoxidans* strain F3B for improving phytoremediation of phenolic pollutants. J Hazard Mater 219:43–49

Holba AG, Dzou IL, Hickey JJ, Franks SG, May SJ, Lenney T (1996) Reservoir geochemistry of South Pass 61 field, Gulf of Mexico: compositional heterogeneities relfecting filling history and biodegradation. Organic Giochemistry 24(12):1179–1198

Jaiswal S, Singh DK, Shukla P (2019) Gene editing and systems biology tools for pesticide bioremediation: a review. Front Microbiol 10:87

Joshi N, Deepali D (2012) Study of ground water quality in and around Sidcul industrial area, Haridwar, Uttarakhand, India. Int J Appl Technol Environ Sanitation 2(2):129–134

Kadiyala V, Spain JC (1998) A two-component monooxygenase catalyzes both the hydroxylation of p-nitrophenol and the oxidative release of nitrite from 4-nitrocatechol in *Bacillus sphaericus* JS905. Appl Environ Microbiol 64(7):2479–2484

Karakaya P, Christodoulatos C, Koutsospyros A, Balas W, Nicolich S, Sidhoum M (2009) Biodegradation of the high explosive hexanitrohexaazaiso-wurtzitane (CL-20). Int J Environ Res Public Health 6(4):1371–1392

Kaushik KA, Dalal SJ, Panwar S (2012) Impact of industrialization on culture of Uttarakhand and its role on career enhancement. VSRD Int J Bus Manag Res 2(4):123

Kavitha GV, Beebi SK (2003) Biodegradation of phenol in a packed bed reactor using peat media. Asian J Microbiol Biotechnol Environ Exp Sci 5(2):157–159

Kawa-Rygielska J, Chmielewska J, Plaskowska E (2007) Effect of raw material quality on fermentation activity of distillery yeast. Polish J Food Nutr Sci 57:275–279

Khan NT (2018) Integration of bioinformatics in bioremediation. Int J Biomed Data Min 7:1000130. https://doi.org/10.4172/2090-4924.1000130

Khanna S, Kumar A (2022) Bioinformatics toward improving bioremediation. In: Arora S, Kumar A, Ogita S, Yau YY (eds) Biotechnological innovations for environmental bioremediation. Springer, Singapore. https://doi.org/10.1007/978-981-16-9001-3_27

Khongkhaem P, Intasiri A, Luepromchai E (2011) Silica-immobilized *Methylobacterium* sp. NP3 and *Acinetobacter* sp. PK1 degrade high concentrations of phenol. Lett Appl Microbiol 52:448–452

Kort MJ (1979) Colour in the sugar industry. In: de Birch GG, Parker KJ (eds) Science and technology. Applied Science, London, pp 97–130

Kumar A, Kumar S, Kumar S (2005) Biodegradation kinetics of phenol and catechol using *Pseudomonas putida* MTCC 1194. Biochem Eng J 22(2):151

Kumari P, Kumar Y (2021) Bioinformatics and computational tools in bioremediation and biodegradation of environmental pollutants. In: Kumar V, Saxena G, Shah MP (eds) Bioremediation for environmental sustainability. Elsevier, pp 421–444

Lawrence M, Huber W, Pages H, Aboyoun P, Carlson M, Gentleman R, Carey VJ (2013) Software for computing and an-notating genomic ranges. PLoS Comput Biol 9:e1003118

Lorenz A, Rylott EL, Strand SE, Bruce NC (2013) Towards engineering degradation of the explosive pollutant hexahydro-1,3,5-trinitro-1,3,5-triazine in the rhizosphere. FEMS Microbiol Lett 340(1):49–54

Madaeni SS, Eslamifard MR (2010) Recycle unit wastewater treatment in petrochemical complex using reverse osmosis process. J Hazard Mater 124:404–409

Madukasi EI, Dai X, He C, Zhou J (2010) Potentials of phototrophic bacteria in treating pharmaceutical wastewater. Int J Environ Sci Technol 7(1):165–174

Margesin R, Fonteyne PA, Redl B (2005) Low-temperature biodegradation of high amounts of phenol by *Rhodococcus* spp. and basidiomycetous yeasts. Res Microbiol 156:68–75

Marimuthu T, Rajendran S, Manivannan M (2013) A review on fungal degradation of textile dye effluent. Acta Chimica Pharmaceutica Indica 3(2):192–200

Maszenan AM, Liu Y, Ng WJ (2011) Bioremediation of wastewater with recalcitrant organic compounds and metals by aerobic granules. Biotechnol Adv 29:111–123

McIlroy S, Saunders AM, Albertsen M (2015) MiDAS: a field guide to the microbes of activated sludge. Database 2015. https://doi.org/10.1093/database/bav062

Mercimek HA, Dincer S, Guzeldag G, Ozsavli A, Matyar F, Arkut A, Kayis F, Ozdenefe MS (2015) Degradation of 2,4,6-trinitrotoluene by *P. aeruginosa* and characterization of some metabolites. Braz J Microbiol 46(1):103–111

Mohammad P, Azarmidokht H, Fatollah M, Mahboubeh B (2006) Application of response surface methodology for optimization of important parameters in decolorizing treated distillery wastewater using *Aspergillus fumigates* UB260. Int Biodeterior Biodegradation 57(4):195

Mohan N, Balasubramanian N, Ahmed Basha C (2007) Electrochemical oxidation of textile wastewater and its reuse. J Hazard Mater B 147:644–651

Mohanty G, Mukherji S (2008) Biodegradation rate of diesel range n-alkanes by bacterial cultures *Exiguobacterium aurantiacum* and *Burkholderia cepacia*. Int Biodeterior Biodegradation 61(3):240–250

Moller S, Pedersen AR, Poulsen LK, Arvin E, Molin S (1996) Activity and three dimensional distribution of toluene-degrading *Pseudomonas putida* in a multispecies biofilm assessed by microscopy. Appl Environ Microbiol 12:4632–4640

Nigam S, Sinha S (2023) Bioinformatics and its contribution to bioremediation and genomics. In: Kumar V, Bilal M, Ferreira LFR, Iqbal HMN (eds) Genomics approach to bioremediation. https://doi.org/10.1002/9781119852131.ch24

Ntuli F, Omoregbe I, Kujipa P, Muzenda E, Belaid M (2009) Characterisation of effluent from textile wet finishing operations, vol 1. WCECS, San Francisco, USA

Pandey RA, Padoley KV, Mukerji SS, Mudliar SN, Vaidya AN, Rajvaidya AS, Subbarao TV (2007) Biotreatment of waste gas containing pyridine in a biofilter. Biores Technol 98:2258

Panikov NS, Sizova MV, Ros D, Christodoulatos C, Balas W, Nicolich S (2007) Biodegradation kinetics of the nitramine explosive CL-20 in soil and microbial cultures. Biodegradation 18(3):317

Pant D, Adholeya A (2007) Identification, ligninolytic enzyme activity and decolorization potential of two fungi isolated from a distillery effluent contaminated site. Water Air Soil Pollut 183:165–176

Pant D, Adholeya A (2010) Development of a novel fungal consortium for the treatment of molasses distillery wastewater. Environmentalist 30:178–182

Qu Y, Ma Q, Deng J, Shen W, Zhang X, He Z, Van Nostr JD, Zhou J, Zhou J (2015) Responses of microbial communities to single walled carbon nano tubes in phenol wastewater treatment system. Environ Sci Technol 49:4627–4635

Rani N, Sangwan P, Joshi M, Sagar A, Bala K (2019) Microbes: a key player in industrial wastewater treatment. In: Shah MP, Rodriguez-Couto S (eds) Microbial wastewater treatment, chap. 5. Elsevier, pp 83–102. https://doi.org/10.1016/B978-0-12-816809-7.00005-1

Rodríguez A, Castrejón-Godínez ML, Sánchez-Salinas E, Mussali-Galante P, Tovar-Sánchez E, Ortiz-Hernández M (2022) Pesticide bioremediation: OMICs technologies for understanding the processes. In: Pesticides bioremediation. Springer, Cham, Switzerland, pp 197–242

Sanghvi G, Thanki A, Pandey S, Singh NK (2020) Engineered bacteria for bioremediation. In: Bioremediation of pollutants. Elsevier, Amsterdam, The Netherlands, pp 359–374

Sangwan P, Celin SM, Hooda L (2015) Response surface methodological approach for optimizing process variables for Biodegradation of 2,4,6-Trinitrotoluene using *Acinetobacter noscomialis*. Eur J Adv Eng Technol 4:51–56

Sar P, Islam E (2012) Metagenomic approaches in microbial bioremediation of metals and radionuclides. In: Microorganisms in environmental management. Springer, Dordrecht, The Netherlands, pp 525–546

Sharma P, Singh SP, Iqbal HM, Tong YW (2022) Omics approaches in bioremediation of environmental contaminants: an integrated approach for environmental safety and sustainability. Environ Res 211:113102

Shim H, Hwang B, Lee SS, Kong SH (2005) Kinetics of BTEX biodegradation by a coculture of *Pseudomonas putida* and *Pseudomonas fluorescens* under hypoxic conditions. Biodegradation 16:319–327

Shokrollahzadeh S, Azizmohseni F, Golmohammad F, Shokouhi H, Khademhaghighat F (2008) Biodegradation potential and bacterial diversity of a petrochemical wastewater treatment plant in Iran. Biores Technol 99(14):6127–6133

Singh D, Geat N, Mehriya M, Rajawat MVS, Prasanna R, Kumar A, Kumari G, Jha MN (2022) Omics (genomics, proteomics, metabolomics, etc.) tools to study the environmental microbiome and bioremediation. In: Waste to energy: prospects and applications. Springer, Singapore, pp 235–260

Sisco E, Najarro M, Samarov D, Lawrence J (2015) Quantifying the degradation of TNT and RDX in a saline environment with and without UV-exposure. Forensic Sci Int 251:124–131

Spina F, Anastasi A, Prigione V, Tigini V, Varese GC (2012) Biological treatment of industrial wastewater: a fungal approach. Chem Eng Trans 27:175–180

Srivastava S, Thakur IS (2007) Evaluation of biosorption potency of *Acinetobacter* sp. for removal of hexavalent chromium from tannery effluent. Iran J Environ Health Sci Eng 5(3):195–200

Suresh A, Abraham J (2018) Bioremediation of hormones from wastewater. In: Hussain CM (ed) Handbook of environmental material management. Springer International Publishing AG

Takeo M, Abe Y, Negoro S, Heiss G (2003) Simultaneous degradation of 4-nitrophenol and picric acid by two different mechanisms of *Rhodococcus* sp. PN1. J Chem Eng Jpn 36(10):1178–1184

Thangaraj K, Kapley A, Purohit HJ (2007) Characterization of diverse *Acinetobacter* isolates for utilization of multiple aromatic compounds. Biores Technol 99:2488–2498

Tripathi M, Singh DN, Vikram S, Singh VS, Kumar S (2018) Metagenomic approach towards bioprospection of novel biomolecule(s) and environmental bioremediation. Ann Res Rev Biol 22:1–12

U.S. Environmental Protection Agency (1997) Profile of the Textile Industry, office of compliance sector notebook project. US Environmental Protection Agency, Washington, D.C., p 40

Vega-Páez JD, Rivas RE, Dussán-Garzón J (2019) High efficiency mercury sorption by dead biomass of *Lysinibacillus sphaercus*—new insights into the treatment of contaminated water. Materials 12:1296

Verma S, Prasad B, Mishra IM (2010) Pretreatment of petrochemical wastewater by coagulation and flocculation and the sludge characteristics. J Hazard Mater 178:1055–1064

Verma S, Kour S, Pathak RK (2021) In silico approaches in bioremediation research and advancements. In: Bioremediation of environmental pollutants, pp 221–238. https://doi.org/10.1007/978-3-030-86169-8_9

Villegas-Plazas M, Sanabria J, Junca H (2019) A composite taxonomical and functional framework of microbiomes under acid mine drainage bioremediation systems. J Environ Manag 251:109581

Wang D, Boukhalfa H, Marina O, Ware DS, Goering TJ, Sun F, Daligault HE, Chi Lo C, Vuyisich M, Starkenburg S (2017) Biostimulation and microbial community profiling reveal insights on RDX transformation in groundwater. Microbiol Open 6(2)

Ward DM, Brock T (1978) Hydrocarbon biodegradation in hypersaline environments. Appl Environ Microbiol 35(2):353–359

Watanabe Y, Sugi R, Tanaka Y, Hayashida S (1982) Enzymatic decolourization of melanoidin by *Coriolus* sp. Agric Biol Chem 46(20):1623–1630

Wilkie AC, Riedesel KJ, Ownes JM (2000) Spillage characterization and anaerobic treatment of ethanol stillage from conventional and cellulosic feedstocks. Biomass Bioenerg 19:63–102

Yang Q, Yang M, Pritsch K, Yediler A, Kettrup A (2003) Decolorisation of synthetic dyes and production of manganese dependent peroxidase by new fungal isolates. Biotech Lett 25:709–713

Yang Q, Liang Y, Zhou T (2008) TNT determination based on its degradation by immobilized HRP with electrochemical sensor. Electrochem Commun 10(8):1176–1179

Yergeau E, Sanschagrin S, Beaumier D, Greer CW (2012) Metagenomic analysis of the bioremediation of diesel-contaminated Canadian high Arctic soils. PLoS ONE 7:e30058

Yeruva DK, Jukuri S, Velvizhi G, Kumar NA, Swamy YV, Mohan VS (2015) Integrating sequencing batch with bioelectrochemical treatment for augmenting remediation efficiency of complex petrochemical wastewater. Biores Technol 188:33–42

Yunusa YR, Umar ZD (2021) Effective microbial bioremediation via the multi-omics approach: an overview of trends, problems and prospects. UMYU J Microbiol Res 6:127–145

Zhang W, Li F, Nie L (2010) Integrating multiple 'omics' analysis for microbial biology: application and methodologies. Microbiology 156:287–301

Zhang T, Shao MF, Ye L (2012) 454 Pyrosequencing reveals bacterial diversity of activated sludge from 14 sewage treatment plants. ISME J 6(6):1137–1147

Zheng Y, Li Y, Long H, Zhao X, Jia K, Li J, Zhang D (2018) *BifA* regulates biofilm development of *Pseudomonas putida* MnB1 as a primary response to H_2O_2 and Mn^{2+}. Front Microbiol 9:1490

Chapter 5
Challenges and Future Perspectives

Abstract Waste production increases due to industrial commodity needs and human population growth. Bioremediation, using synthetic microorganisms and GMOs, is being studied but risks genetic pollution. Waste management is crucial for sustainable development, but progress is limited. Rapid urbanization and industrialization increase waste management demands, emphasizing proactive measures like separation, recycling, and recovery. Biotechnology and biodegradation can enhance efficiency. Long-term strategies are essential for adapting to changing lifestyles.

Keywords Bioremediation · Microorganisms · Biodegradation · Genetically modified organism · Bioaugmentation · Waste management · Pollution management · Genetic manipulation

5.1 Challenges

The growth of the human population results in amplified demands for industrial commodities, which subsequently escalate the volume of waste produced through manufacturing processes. Bioremediation is recognized as an appealing green solution, largely because of its suitability and affordability (Azubuike et al. 2020; Mondal and Palit 2019; Parmar et al. 2021; Satyanarayana et al. 2012; Vishwakarma et al. 2020). But, like any process, it is not without its challenges and limitations. A primary challenge is the scaling-up of operations from laboratory settings to real-world applications, along with the necessity to reduce the time needed for complete detoxification. Research efforts are now directed toward exploring the role of genetically modified organisms (GMOs) and synthetically engineered microbes to optimize performance under specific conditions. Nonetheless, these advancements could lead to significant genetic pollution and may result in unpredictable effects if not handled with care. In this regard, GMOs designed to self-eliminate after fulfilling their purpose could be particularly beneficial (Macaulay and Rees 2014; Ramırez-Garcıa et al. 2019).

P. Bajpai, *Developments in Microbial Bioremediation*,
SpringerBriefs in Molecular Science, https://doi.org/10.1007/978-3-031-78319-7_5

Insufficient expertise in various scientific disciplines related to bioremediation constitutes a major issue. It is imperative to have a solid foundation in fields such as microbiology, genomics, structural and molecular biology, hydrology, and chemistry before embarking on the design of any remediation initiatives. Another significant challenge lies in determining the specific pathways through which microbes operate to rehabilitate polluted environments. A comprehensive examination of the chemical processes and the interactions among pollutants, native microorganisms, and the intricate systems present at contaminated locations is crucial. Furthermore, the lack of coordinated research initiatives in this field poses an additional obstacle. The effective implementation of bioremediation strategies necessitates the joint efforts of biochemists, microbiologists, biotechnologists, soil scientists, and engineers. Promoting multidisciplinary study and the sharing of scientific knowledge can enhance the efficacy of these techniques. Unfortunately, the field of bioremediation often does not receive adequate attention, primarily due to its limited ability to generate significant revenue. Both government and private sectors find it challenging to achieve desired financial returns from this area, leading to a delay in progress as materialistic incentives remain minimal (Ramırez-Garcıa et al. 2019; Macaulay and Rees 2014).

A significant limitation of bioremediation is that not all compounds possess biodegradability, which restricts the effectiveness of the bioremediation process. In some cases, the processing and breakdown of a biodegradable material can result in the formation of harmful byproducts, despite the material's inherent biodegradability. Furthermore, a particular bacterial strain that demonstrates effectiveness in one environment may not yield satisfactory results in another due to localized limiting factors. The intricacy of biological systems is shaped by the metabolic activities of microorganisms and the specific attributes of the pollutants involved and the accessibility of vital nutrients. The requirement for soil excavation and the construction of specialized site layouts contribute to the time-consuming and labor-intensive nature of the process. Bioremediation is generally performed in subterranean and secluded locations, away from populated areas. However, there may be noise and disturbances associated with the usage of large gear and pumps. Furthermore, questions concerning the impact of particular bacterial cultures on the native microflora may arise due to ethical concerns around their use (Ramırez-Garcıa et al. 2019).

Bioremediation has emerged as a crucial technique in the realm of environmental science for tackling the contamination of aquifers and soils. The capacity of specific microorganisms to decompose environmental pollutants is now well-documented. However, challenges arise when attempting to translate this microbial capability from laboratory settings to real-world applications. Three primary limitations have been identified that hinder the widespread adoption of bioremediation techniques. These are:

- Insufficient understanding of microbial behavior in field conditions,
- Difficulties in microbial management and stimulation,
- Challenges in ensuring effective interaction between microbes and contaminants.

Ongoing research by the scientific community in this field has made notable advancements in refining existing methodologies. Researchers globally are exploring innovative engineering solutions to enhance microbial stimulation. An important case in point is the recent innovation in gas-sparging, which has markedly improved the aerobic breakdown of petroleum hydrocarbons. The rapid evolution of bioaugmentation technologies is anticipated to enhance control over genetic manipulation of microorganisms, thereby addressing challenges related to microbiological factors. A thorough comprehension of biotransformation, encompassing both ecological and genetic dimensions, facilitates the creation of advanced bioremediation methods. This expertise can be utilized to formulate strategies for the treatment of hazardous pollutants, including chlorinated solvents and polychlorinated biphenyls, which were previously considered resistant to degradation. Many contaminants evade microbial degradation primarily due to the inability of microbes to effectively interact with or adhere to their surfaces. Consequently, new methodologies are being explored to enhance microbial bioavailability, thereby increasing the efficacy of bioremediation efforts. Methods such as waste solubilization via heat injection, which encompasses the use of steam, hot air, or heated water flushing, alongside high-pressure subsurface matrix fracturing and the application of surfactants, exemplify some of the innovative strategies aimed at enhancing existing technologies.

A comprehensive understanding of bioremediation necessitates not only the identification of effective and practical strategies but also the development of methods to assess the performance and effectiveness of existing techniques. Consequently, various protocols have been established that focus on three fundamental components:

(1) Recording the comprehensive decrease of pollutants at the impacted location;
(2) Assessing the biodegradation potential of microorganisms obtained from the site via laboratory examinations;
(3) Considering additional informational factors that emphasize the operational success of the methods applied in the field.

The main goal is to evaluate the practicality of theoretically suggested solutions to guarantee the fulfillment of cleanup objectives. There is a growing tendency among researchers to develop revised protocols based on molecular biology, which are anticipated to be trustworthy, quick, and affordable. Additionally, the in situ characterization of physicochemical parameters shows potential and could revolutionize field assessment techniques in the future.

Bioremediation is gaining recognition as an effective approach for environmental restoration, largely because it can meet remediation goals in a cost-efficient manner while reducing the likelihood of contaminant spread to surrounding areas. Despite the advantages of bioremediation, the need for meticulous management of microbial agents and continuous monitoring at subsurface levels leads some clients to hesitate in selecting this method for management of pollutants and waste. However, the future appears promising with the potential for novel methods that can address current challenges and steer society toward a more sustainable and cleaner environment.

5.2   Future Perspectives

Contemporary society relies heavily on a diverse range of industries. Across the globe, the industrial products produced have transformed the lives of millions. However, a significant concern arises from the quality of wastewater generated by these industries, necessitating timely assessment and characterization. In many developing nations, untreated wastewater is frequently discharged, leading to environmental degradation. The bioremediation of wastewater through various microorganisms has garnered the interest of researchers globally and presents considerable potential for the future. In contrast to traditional wastewater treatment methods, microbial degradation plays a crucial role in yielding non-toxic end products and is a more cost-effective solution.

Numerous research initiatives have been undertaken to explore the application of microbial enzymes in the bioremediation of waste materials (Vishwakarma et al. 2020; Ayilara and Babalola 2023; Bala et al. 2022; Mehta et al. 2024). Nevertheless, enhancing these processes is crucial for promoting a safer and more sustainable environment. There is an immediate need for additional research aimed at identifying new microbes capable of efficiently and swiftly bioremediating various pollutants, particularly those generated by industrial activities. It is possible that these newly discovered microbes and their enzymes possess superior capabilities for pollutant degradation compared to those currently in use. It is imperative to undertake further research aimed at establishing swift detection techniques that can assess the advancement of biodegradation or verify the total decomposition of environmental pollutants. Moreover, current microorganisms employed in bioremediation could be genetically modified to enhance their enzyme production, thus increasing their efficiency in biodegradation. Employing a diverse microbial consortium, rather than relying on a single type, may yield better results in bioremediation by incorporating various organisms that can utilize different substrates, ultimately enhancing the rate of microbial degradation.

The practice of bioremediation has become a favored approach for addressing contamination in aquifers and soil. The established capability of certain microorganisms to break down environmental pollutants is well recognized. Nonetheless, the challenge lies in transferring this microbial efficacy from controlled laboratory environments to practical field applications (Mondal et al. 2023).

While microbial bioremediation has demonstrated considerable promise in the remediation of waste materials, there remain several areas that necessitate additional research and advancement to promote a safer and more sustainable environment. The discovery of novel microorganisms capable of effectively bioremediating a range of contaminants, particularly those originating from industrial activities, is essential for improving the removal of pollutants from ecosystems. Microbial glycoconjugates derived from *Scedosporium* sp. and *Acinetobacter* sp. may inherently possess the capacity to remediate these contaminants. The identification and characterization of their enzymes could yield significant insights into their biodegradation processes, thereby informing the creation of more effective bioremediation techniques. Furthermore, the establishment of rapid and dependable detection methods is vital for tracking the progress of bioremediation efforts and verifying the complete degradation of pollutants in the environment. Advanced detection methodologies allow for precise evaluation of the efficacy of bioremediation initiatives and enable timely modifications when required (Visvanathan et al. 2024).

The genetic modification of microorganisms currently employed in bioremediation has the potential to enhance their capabilities for biodegradation. By engineering these organisms to produce increased quantities of enzymes or to specifically target certain pollutants, their overall efficiency and effectiveness in breaking down contaminants can be markedly improved, facilitating customized and focused bioremediation strategies. Rather than depending on a singular microbial consortium, the integration of various microbial consortia may yield a more effective bioremediation approach. This method introduces a diverse array of organisms that can utilize different substrates, thereby augmenting the overall rate of microbial biodegradation and improving the efficiency and adaptability of bioremediation processes. Although microbes are frequently utilized to decompose organic substrates, the remediation of persistent inorganic pollutants continues to pose a significant challenge. Therefore, intensified research efforts should be aimed at identifying and developing microorganisms capable of degrading inorganic pollutants, thus enabling a more holistic approach to bioremediation. The potential role of microbes and their enzymes in the bioremediation of nuclear waste is an area that deserves thorough investigation.

Additional research and development are essential to assess the viability and efficacy of microbial bioremediation in this particular context. While microbes play a crucial role in bioremediation, they can occasionally contribute to pollution. To avoid adverse effects, such as algal blooms resulting from microbial biostimulation, it is imperative to establish effective strategies and regulations to address the potential risks linked to the introduction of microbes into designated environments. Regulatory agencies should be created to oversee and evaluate these risks. Furthermore, enhancing awareness and encouraging the use of microbial degradation techniques are vital. It is important for both policymakers and the general public to be informed about the benefits and safety of microbial biodegradation in comparison to traditional methods, thereby facilitating more informed choices regarding the most effective and environmentally sustainable remediation strategies.

The direct access to environmental microorganisms has significantly revolutionized the discipline of microbial ecology, allowing researchers to investigate and understand microbial communities across various ecosystems. Nevertheless, in the context of bioremediation, the capabilities of these technologies remain largely nascent. Although bioremediation methods exhibit considerable potential, there is still a substantial amount to discover regarding their efficacy and practical implementation. A key benefit of accessing environmental microorganisms is the opportunity to obtain critical and timely data for the management and remediation of polluted environments. By examining the microbial communities present, researchers can gain insights into the feasibility of bioremediation and formulate more effective cleanup strategies. Furthermore, these technologies may significantly advance the identification of novel catabolic processes, which are essential for the degradation of complex compounds into simpler forms in the context of bioremediation.

Through the examination of the genetic composition and metabolic functions of environmental microorganisms, scientists can identify novel pathways and enzymes that may be utilized for bioremediation efforts. These molecular tools are crucial as they adhere to the tenets of sustainable development. By capitalizing on the inherent abilities of microorganisms, bioremediation can evolve into a more eco-friendly strategy for tackling pollution, emphasizing the necessity of sustainable solutions that reduce environmental impact while effectively addressing contaminated areas. The exploration of environmental microorganisms has transformed the field of microbial ecology and presents significant opportunities for bioremediation. These advancements not only yield essential insights for the management and restoration of polluted environments but also have the potential to reveal new catabolic pathways, thereby enhancing the sustainability and environmental compatibility of bioremediation practices in accordance with sustainable development principles.

While bioremediation holds great potential, it is crucial to handle microbes with care and maintain continuous monitoring of their activity in subsurface environments. The future of bioremediation will depend on the development of rapid technologies that address current challenges and contribute to a cleaner, greener planet (Khan 2024).

References

Ayilara MS, Babalola OO (2023) Bioremediation of environmental wastes: the role of microorganisms. Front Agron 5:1183691. https://doi.org/10.3389/fagro.2023.1183691

Azubuike CC, Chikere CB, Okpokwasili GC (2020) Bioremediation: an eco-friendly sustainable technology for environmental management. Bioremediation of industrial waste for environmental safety. Springer, Singapore, pp 19–39

Bala S, Garg D, Thirumalesh BV, Sharma M, Sridhar K, Inbaraj BS, Tripathi M (2022) Recent strategies for bioremediation of emerging pollutants: a review for a green and sustainable environment. Toxics 10(8):484. https://doi.org/10.3390/toxics10080484

Khan MS (2024) Applications of bioremediation in biomedical waste management: current and future prospects. Environ Sci Braz Arch Biol Technol 67:e24230161. https://doi.org/10.1590/1678-4324-2024230161

Macaulay BM, Rees D (2014) Bioremediation of oil spills: a review of challenges for research advancement. Ann Environ Sci 8:9–37

Mehta S, Prasad ME, Upadhye VJ, Goswami S, Singh P (2024) Enhancing efficacy of microbial bioremediation by intervention of nanotechnology and metabolic engineering: a review. J Appl Nat Sci 16(2):741–751. https://doi.org/10.31018/jans.v16i2.5520

Mondal S, Palit D (2019) Effective role of microorganism in waste management and environmental sustainability. In: Sustainable agriculture, forest and environmental management. Springer, Singapore, pp 485–51

Mondal S, Mukherjee SK, Hossain ST (2023) Exploration of plant growth promoting rhizobacteria (PGPRs) for heavy metal bioremediation and environmental sustainability: recent advances and future prospects. In: Modern approaches in waste bioremediation: environmental microbiology, pp 29–55. https://doi.org/10.1007/978-3-031-24086-7_3

Parmar S, Daki S, Bhattacharya S Shrivastav A (2021) Microorganism: an ecofriendly tool for waste management and environmental safety. In: Shah MP, Rodriguez-Couto S, Kapoor RT (eds) Development in wastewater treatment research and processes, Chap. 8. Elsevier, pp 175–193. ISBN 9780323856577. https://doi.org/10.1016/B978-0-323-85657-7.00001-8.

Ramırez-Garcıa R, Gohil N, Singh V (2019) Recent advances, challenges, and opportunities in bioremediation of hazardous materials. In: Pandey VC, Bauddh K (eds) Phytomanagement of polluted sites. Elsevier, Amsterdam, pp 517_56

Satyanarayana T, Johri BN, Prakash A (eds) (2012) Microorganisms in environmental management: microbes and environment. Springr Science & Business Media

Vishwakarma GS, Bhattacharjee G, Gohil N, Singh V (2020) Current status, challenges and future of bioremediation. In: Pandey VC, Singh V (eds) Bioremediation of pollutants. Elsevier, pp 403–415. https://doi.org/10.1016/B978-0-12-819025-8.00020-X

Visvanathan P, Kirubakaran D, Selvam K, Manimegalai P, Rajkumar M, Vasantharaj K (2024) Recent strategies for natural bioremediation of emerging pollutants: development of a green and sustainable environment. Bioremediat Emerg Contam Water 2:1–19. https://doi.org/10.1021/bk-2024-1476.ch001

Index

A

Acaligenes sp., 45
Acenitobacter noscomialis, 87
Acetobacter acetii, 89
Achromobacter sp., 45, 85
Acid insoluble polyphosphate, 55
Acidithiobacillus ferrooxidans, 27
Acidobacteria, 52
Acinetobacter, 15, 16, 28, 34, 35, 37, 45,
  46, 85, 86, 91, 106
Acinetobacter junii TYRH47, 35
Acinetobacter sp. BRS156
  Acremonium sp., 48
Actinobacteria, 86
Activated sludge, 36, 51, 52, 56, 62, 66, 82,
  84, 85
Acute conjunctivitis, 12
Adenoviruses, 12
Adsorption, 47, 48, 85, 88, 89
Advanced treatment, 35, 67
Advanced wastewater treatment, 67
Aerobic, 10, 12, 25, 29, 45, 47, 51–53, 56,
  66, 82, 85, 87, 90, 105
Agricultural products, 88
Agricultural waste, 2, 7, 12
Agrochemicals, 17
Air pollutants, 29
Alcanivorax, 15
Algae, 3, 10, 12, 14–17, 33–38, 45, 47,
  50–52, 54, 55, 57–59, 70, 84, 85, 89,
  90
Algal-bacterial consortium, 34
Alginate, 69
Alicaligens sp., 89
Aliphatic hydrocarbons, 90
Alkanes, 91

benzene, 16, 91
  toluene, 16, 91
Aminodinitrotoluene, 87
4-aminodinitrotoluene, 87
Ammonia, 28, 30, 47, 54, 56–58, 69, 70, 91
Amoebae, 11, 50
Anabaena oryzae, 45
Anabaena sp., 47
Anacystis, 12
Anaerobic, 10, 25, 26, 44, 45, 47, 56, 63,
  66, 82, 87, 91
Anaerobic film reactors, 82
Anaerobic filters, 82
Anaerobic processes, 53, 70, 90
Anaerobic sludge reactors, 82
Anaerobiosis, 91
Anaerolineales, 84
Animal waste, 9
Antibiotics, 34, 44, 49, 82
Aphanocapsu sp., 45
Arthrobacter sp., 45, 85
Arthropods crustaceans, 12
Artificial wetlands, 65, 66
Aseptic meningitis, 12
Aspergillus, 16, 28, 35–37, 48, 83, 86
Aspergillus foetidus, 85
Aspergillus niger, 37, 49, 89
Aspergillus oryzae
  Aspergillus parasitica, 48
Aspergillus ustus, 85
Autotrophic, 30, 36, 45, 54, 56
Azo dyes, 6, 16, 34
Azotobacter vinelandii, 69

P. Bajpai, *Developments in Microbial Bioremediation*,
SpringerBriefs in Molecular Science, https://doi.org/10.1007/978-3-031-78319-7